A Concise Introduction to Robot Programming with ROS 2

A Concise Introduction to Robot Programming with ROS 2 provides the reader with the concepts and tools necessary to bring a robot to life through programming. It will equip the reader with the skills necessary to undertake projects with ROS 2, the new version of ROS. It is not necessary to have previous experience with ROS 2, as it will describe its concepts, tools, and methodologies from the beginning.

- Uses the two programming languages officially supported in ROS 2 (C++, mainly, and Python)
- Approaches ROS 2 from three different but complementary dimensions: the Community, Computation Graph, and the Workspace
- Includes a complete simulated robot, development and testing strategies, Behavior Trees, and Nav2 description, setup, and use
- A GitHub repository with code to assist readers

It will appeal to motivated engineering students, engineers, and professionals working with robot programming.

Francisco Martín Rico is a Full Professor at the Rey Juan Carlos University, Madrid, leading the Intelligent Robotics Lab. He teaches subjects such as Software Architectures for Robots, Mobile Robotics, and Planning and Cognitive Systems. He is a reputable member of the ROS Community, authoring PlanSys2, the symbolic planning reference software for ROS 2, and many other packages such as YAETS and Cascade Lifecycle, and contributing to reference packages like Nav2. He was awarded in 2022 with the Best ROS Developer award at the ROS Developer Day and has been a member of the ROS 2 Technical Steering Committee.

A Concise Introduction to Robot Programming with ROS 2

Second Edition

Francisco Martín Rico

CRC Press
Taylor & Francis Group
Boca Raton London New York

CRC Press is an imprint of the
Taylor & Francis Group, an **informa** business

A CHAPMAN & HALL BOOK

Front cover image: Francisco Martín Rico

Second edition published 2025
by CRC Press
2385 NW Executive Center Drive, Suite 320, Boca Raton FL 33431

and by CRC Press
4 Park Square, Milton Park, Abingdon, Oxon, OX14 4RN

CRC Press is an imprint of Taylor & Francis Group, LLC

© 2025 Francisco Martín Rico

First edition published by CRC Press 2022

Library of Congress Cataloging-in-Publication Data
Names: Rico, Francisco Martín author
Title: A concise introduction to robot programming with ROS2 / Francisco Martín Rico.
Description: Second edition. | Boca Raton, FL : CRC Press, 2025. | Includes bibliographical references and index. |
Identifiers: LCCN 2024061328 (print) | LCCN 2024061329 (ebook) | ISBN 9781032851501 hbk | ISBN 9781032851488 pbk | ISBN 9781003516798 ebk
Subjects: LCSH: Robots--Programming
Classification: LCC TJ211.45 .R53 2025 (print) | LCC TJ211.45 (ebook) |
DDC 629.8/92--dc23/eng/20250326
LC record available at https://lccn.loc.gov/2024061328
LC ebook record available at https://lccn.loc.gov/2024061329

ISBN: 978-1-032-85150-1 (hbk)
ISBN: 978-1-032-85148-8 (pbk)
ISBN: 978-1-003-51679-8 (ebk)

DOI: 10.1201/9781003516798

Typeset in Latin Modern font
by KnowledgeWorks Global Ltd.

Publisher's note: This book has been prepared from camera-ready copy provided by the author.

Contents

List of Figures

Acknowledgements

I would like to thank the people who have made this book possible.

First of all, to Steve Macenski and Michele Collendise, who counted on me for the book "Making Robots Work".

Secondly, to Fran and Vicente, professors and researchers. They introduced me to ROS and Robotics, respectively, and have been the best journey companions through the world of Science and the University.

I want to make a special mention to the members of the ROS Community who have dedicated their time and effort to improve this book, sending comments and revisions to its first published version. Andrej Orsula, Sam Pfeiffer, and Zahi Kakish sent me extensive, high-quality reviews that greatly enhanced the text quality. Varun Vivek Vennavalli, Brian Hope, and Jorge Turiel also contributed with corrections.

I also want to thank some members of my group who helped me to make this second edition reality. Thanks Juan Carlos Manzanares and Juan Sebàstian Cely for your help during summer'24 to migrate the simulation and projects from Gazebo Classic to Gazebo Harmonic for this second edition. I extend my gratitude to Rodrigo Perez and Jose Miguel Guerrero for dedicating their time and effort to reviewing the two additional chapters in this second edition. Thanks to Jesus M. Gonzalez Barahona and Gregorio Robles, two world-renowned experts in Free Software, for guiding me and reviewing the section about software licenses.

Lastly, I would like to thank my parents for having gone all this way to make my education possible, with all the effort it has meant for them.

My final words of gratitude are reserved for the most important person behind this book. I want to express my heartfelt gratitude to Marta for her patience, affection, and love throughout the writing of this book, which has taken so many hours away from her husband. I love you, my dear.

Foreword by Esteve Fernandez

ONe of the most beautiful aspects of free and open-source software is how it connects people from different backgrounds and places. ROS is no different. ROS can be seen as a framework, a protocol or an ecosystem, but for some people—including me—, it's first and foremost a community. Everyone's journey with ROS is unique, and mine started almost 15 years ago.

In 2012, Fluidinfo, the company my good friend Terry Jones and I started, had run out of money. Though the process of winding down the business was slow, it still seemed like an abrupt change when we could no longer work on FluidDB. Suddenly, I no longer had a job, but then I remembered why I got into computers, and that was robotics. I was very lucky, because I found a master's program at Universitat Pompeu Fabra that still accepted applications. If Fluidinfo had shut down a week later, it would have been a very different story. I enrolled in the program to delay the decision of what to do in the future, and just enjoy learning about robotics for a year. Like many other students and researchers at the time, I had the experience of a robotics department with their own homemade robotics framework, with all the pros and cons that involved. I knew there had to be a different way of programming robots, one where certain problems were already solved.

And that's how I found ROS 1 (or simply ROS, at the time). Of course, it had its quirks, but its design made a lot more sense in my mind, as someone whose background was in distributed systems. As a proof of concept to see if I understood the internals of ROS, I created a ROS library for Twisted (remember, async Python didn't exist yet), a framework I was quite familiar with. Happy with the result, I went to my first ROSCon to present it as a lightning talk. A week beforehand I had applied for the Google Summer of Code to work on CloudSim at the Open Source Robotics Foundation, so I was excited to have the chance to meet many of the people at OSRF in person at ROSCon. The talk went well, and I went back to Spain to finish my master's. A few weeks later I was accepted at Google Summer of Code. After Google Summer of Code ended, Brian Gerkey, the CEO of OSRF, sent me an email asking if I'd be interested in interviewing for a job. Of course I said yes! I flew to California for the interview, spent two weeks there working alongside the ROS team and on the way back I got an email with a job offer. I honestly couldn't believe it, it felt like a joke, and I was more than happy to accept! There weren't that many people in the core team: William Woodall, Tully Foote, Morgan Quigley, Jackie Kay, Dirk Thomas and me. We maintained all the essential ROS packages and more, until one day, we talked about creating a successor to ROS that would incorporate all the knowledge the community had acquired over the years. We wanted to create something that would fit their use cases much better, the spark of an idea that would eventually become ROS 2.

Unsurprisingly, ROS 2 started with a new build system, ament, the successor to ROS 1's catkin. Both names came as a nod to the place where it all started, Willow Garage.

Early designs for ROS 2 used ZeroMQ, our own custom protocol, and many others, but DDS seemed to be the one that best fit the specs. DDS was being used in several mission-critical cases, had multiple implementations and was backed by a consortium of companies. Moreover, from the technical side, it had everything we needed in one package: discovery, serialization, pubsub, and services. The early days of ROS 2 were so much fun, but also we knew how important it was for the community to be able to continue programming their robots with ROS. We wanted to show that the break from ROS 1 was necessary in order to implement the vision of a robotics framework that would work for the next generation of robotics. Because of that, we named early ROS 2 Alphas after tools and objects that kept things together (e.g., Alpha 1 was Anchor, Alpha 2 Baling wire, etc.), as we wanted to hold the community together while ROS 2 broke away from ROS 1. ROS 1 was so disruptive because of how it was developed in the open, the same way that developing an operating system kernel in the open was completely unheard of before the Linux kernel. ROS 2, meanwhile, represents the next stage, where robotics has matured enough to reach mass adoption.

After three years at OSRF, I wanted to move on to my next challenge, but I wasn't yet sure what that might be. So I moved to Madrid, where I met Francisco Martín Rico (known to most of us as "Paco"), and we quickly became friends. He was using ROS 1 in his research, but he quickly saw that ROS 2 was going to be the future. We ported ROS 2 to Aldebaran's Pepper robot, to Microsoft's Hololens, and used it in a bunch of other interesting projects. I moved to Paris after Madrid, and although I've worked on many other projects with Paco since, the thing I'm most proud of is our friendship after so many years.

At the last ROSCon in Odense, someone asked me how familiar I was with ROS 2, and I replied, "a bit". It is strangely heartwarming that people are using a software you co-created without even knowing who you are, and that the software has grown a life of its own. This book is proof of that.

I hope you enjoy this book as much as I enjoyed writing the foreword for it, and I hope it helps you in your journey with ROS 2. I can think of no one better to guide you on your journey through ROS 2 than Paco.

Esteve Fernandez
Robotics Engineer, ROS 2 co-author

Preface to the Second Edition

I T has been almost four years since I embarked on this project. Initially, this book was just a chapter for the book "Making Robots Work", which, back in March 2021, was being promoted by Michele Colledanchise and Steve Macenski. It was an ambitious project that originated from a call for authors on ROS Discourse and social media. Each chapter was meant to focus on a different aspect of robotics: Making Robots See, Making Robots Navigate, and so on, covering all the capabilities and facets a robot could perform.

Authors were asked to submit a draft chapter, and I chose to write the introductory chapter, which provided an overview of ROS 2 and set up the simulator to be used throughout the rest of the book. At the end of April, I submitted my draft, and a month later, I was informed that my proposal had been accepted, with a deadline of September to finalize the final chapter. That summer, I worked extremely hard and delivered a 72-page chapter that condensed everything someone needed to know about ROS 2 before tackling the more advanced content in the following chapters. Unfortunately, at the end of October, Michele and Steve informed me that the project had been canceled. It seemed that I was practically the only one who had delivered something close to their vision on time. I was devastated, but they suggested I talk to the editor, Randi Slack, to see if my work could still be put to good use. She proposed expanding it to 100 pages. By March 2022, I had completed the book, which ended up being 251 pages—although almost 100 of those pages consisted of the full source code of the examples. I even created the book's cover using draw.io, relying on simple geometric shapes.

I uploaded the source code of the projects, designed for the Foxy distribution, to a GitHub repository that I have continued to maintain across subsequent distributions (Galactic, Humble, and Iron). The repository also includes slides for the chapters, all openly accessible.

The book was published in the fall of 2022, two years ago now. During this time, its reception within the ROS Community has been spectacular. Many people sent me photos of the book through social media when it arrived, and I have met numerous individuals who have told me that my book greatly helped them in their ROS training. Every time someone asks me to sign their copy, I feel a mix of embarrassment and pride. I believe I have helped many people, and for that, I feel genuinely happy. Sometimes, I experience imposter syndrome when I see the impact and expectations that my work has on others—not just this book, but also all the packages I have created and maintain. I understand that this feeling is normal and that, ironically, only true impostors lack this syndrome.

A few months after the book was published, I invited readers to send me any typos they found, and many people enthusiastically contributed. Unfortunately, these corrections never made it to the first edition but have been incorporated into this second edition. Another milestone for this book came when the publisher informed me it would also be translated into Chinese. I am not sure what became of that, but I still hope to one day see the cover written with sinograms. The most significant milestone, however, was when the publisher proposed that I write this second edition in February 2024.

In this second edition, I decided—without it being strictly necessary—to remove all source code and replace it with two outstanding chapters: Delving deeper into ROS 2 and how to contribute to the ROS ecosystem. In the first chapter, I explored the intricacies of ROS 2's execution model and wrote extensively about real-time systems, drawing heavily from the incredible workshop I attended at ROSCon in New Orleans. For this, I even developed a simple tool to trace program execution. In the second chapter, I focus on the ROS Community, discussing how to contribute to existing projects and how to lead your own, covering critical aspects such as software licensing and using Git the way it is used for ROS project contributions. Beyond these two new chapters, I migrated everything to the Jazzy distribution, which was a significant undertaking, as it required switching from Gazebo Classic to Gazebo Harmonic. I was determined to keep using the Tiago robot because I needed a robot with a neck for the visual attention chapter. While the rest of the book has been thoroughly reviewed, it remains fundamentally the same as the previous edition. I have changed the cover color to ensure that when you purchase the book, there will be no doubt that you are acquiring this second edition, which is a significant improvement over the previous one.

Lastly, I cannot express enough how thrilled I was that my friend Esteve agreed to write the foreword. It is pure gold, in my opinion.

I did not write this book to get rich, but it is evident that it has brought me many benefits. It is probably one of the factors that contributed to my accreditation as a Full Professor in Spain, a requirement to access positions in the prestigious public universities of this great country. Perhaps it also played a role in making me known to the ROS Community, giving me the honor of representing it on the ROS 2 Technical Steering Committee for two years. Who knows what the future holds?

Before concluding these notes on the second edition, I want to leave a word of caution to readers: *knowing ROS does not mean knowing Robotics*. ROS is a powerful tool for developing robotic applications, and you must master it to the point where your lack of expertise with it never limits your ability to achieve your goals in a robotic project. However, it is only the first step toward becoming a great professional in Robotics.

Enjoy reading this book!

Francisco Martín Rico
Full Professor on Robotics

## CHAPTER 1

Introduction

R OBOTS must be programmed to be useful. It is useless for a robot to be a mechanical prodigy without providing it with software that processes the information from the sensors to send the correct commands to the actuators to fulfill the mission for which it was created. This chapter introduces the middlewares for programming robots and, in particular, *ROS 2* [1], which will be the one used in this book.

First of all, nobody starts programming a robot from scratch. Robot software is very complex since we have to face the problem that a robot performs tasks in a real, dynamic, and sometimes unpredictable world. It also must deal with a wide variety of models and types of sensors and actuators. Implementing the necessary drivers or adapting to new hardware components is a titanic effort doomed to failure.

Middleware is a layer of software between the operating system and user applications to carry out the programming of applications in some domains. Middleware usually contains more than libraries, including development and monitoring tools and a development methodology. Figure 1.1 shows a schematic of a system that includes middleware for developing applications for robots.

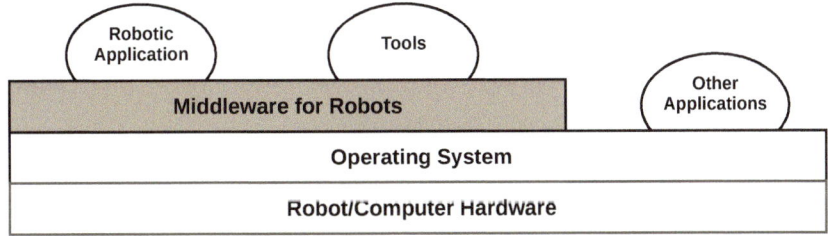

Figure 1.1: Representation of software layers in a Robot.

Robot programming middlewares provide drivers, libraries, and methodologies. They also usually offer development, integration, execution, and monitoring tools. Throughout the history of Robotics, a multitude of robot programming middlewares have emerged. Few of them have survived the robot for which they were designed or have expanded from the laboratories where they were implemented. There are notable examples (YARP [2], Carmen [3], Player/Stage [4], etc.), although without a doubt the most successful in the last decade has been ROS [5], which is currently

DOI: 10.1201/9781003516798-1

considered a standard in the world of robot programming. Technically, there are similarities between the different middlewares: most are based on open source, many provide communication mechanisms for distributed components, compilation systems, monitoring tools, etc. The big difference is the ROS developers community around the world. There are also leading companies, international organizations, and universities worldwide in this community, providing a vast collection of software, drivers, documentation, or questions already resolved to almost any problem that may arise. Robotics can be defined as "the Art of Integration", and ROS offers a lot of software to integrate, as well as the tools to do so.

This book will provide the skills necessary to undertake projects in ROS 2, the new version of ROS. It is unnecessary to have previous experience in ROS 2 since we will describe its concepts, tools, and methodologies from the beginning without the need for previous experience. We will assume average Linux and programming skills. We will use the two programming languages officially supported in ROS 2 (C++ and Python), which coincide with the languages most used in general in Robotics.

1.1   ROS 2 OVERVIEW

The meaning of the acronym ROS is *Robot Operating System*. It is not an operating system that replaces Linux or Windows but a middleware[1] that increases the system's capabilities to develop Robotic applications. The number *2* indicates that it is the second generation of this middleware. The reader who already knows the first version of ROS (sometimes referred to as ROS 1) will find many similar concepts, and there are already several teaching resources[2] for the ROS 1 programmer who lands on ROS 2. In this book, we will assume no previous knowledge of ROS. It will be more and more common for this to happen, as there are now more and more reasons to learn ROS 2 directly instead of going through ROS 1 first.

Also, there are already some excellent official ROS 2 tutorials, so the approach in this book is intended to be different. The description will be complete and with a methodology oriented to developing robotic applications that make the robot do something "smart", from robotic engineer to robotic engineer, emphasizing essential issues that come from experience in the development of software in robots. It will not hurt for the reader to explore the tutorials available to complete their training and fill in the gaps that do not fit in this book:

- Official ROS 2 tutorials: `https://docs.ros.org/en/jazzy/Tutorials.html`

- The Robotics Back-End tutorials: `https://roboticsbackend.com/category/ros2`

- ROS 2 for ROS developers: `https://github.com/fmrico/ros_to_ros2_talk_examples`

---

[1]ROS is not defined as a 100% middleware (`https://answers.ros.org/question/12230/what-is-ros-exactly-middleware-framework-operating-system/`), it is a mixture of middleware, framework, and meta operating system.

[2]`https://github.com/fmrico/ros_to_ros2_talk_examples`

The starting point is a Linux Ubuntu 24.04 LTS system installed on a computer with an AMD64-bit architecture, the most extended one in a personal laptop or desktop computer. The Linux distribution is relevant since ROS 2 is organized in *distributions*. A distribution is a collection of libraries, tools, and applications whose versions are verified to work together correctly. Each distribution has a name and is linked to a version of Ubuntu. The software in a distribution is also guaranteed to work correctly with the software version present on the system. It is possible to use another Linux distribution (Ubuntu, Fedora, Red Hat ...), but the reference is Ubuntu. ROS 2 also works on Windows and Mac, but this book focuses on Linux development. We will use the ROS 2 Jazzy Jalisco version, which corresponds to Ubuntu 24.04.

In this book, we will approach ROS 2 from three different but complementary dimensions:

- **The Community**: The ROS community is a fundamental element when developing applications for robots with this middleware. In addition to providing technical documentation, there is a vast community of developers who contribute with their own applications and utilities through public repositories, to which other developers can contribute. Another member of the community may have already developed something you need.

- **Computation Graph**: The Computational Graph is a running ROS 2 application. This graph is made up of nodes and arcs. The *Node*, the primary computing unit in ROS 2, can collaborate with other nodes using several different communication paradigms to compose a ROS 2 application. This dimension also addresses the monitoring tools, which are also nodes that are inserted in this graph.

- **The Workspace**: The Workspace is the set of software installed on the robot or computer and the programs that the user develops. In contrast to the Computational Graph, which has a dynamic nature, the Workspace is static. This dimension also addresses the development tools to build the elements of the Computational Graph.

1.1.1 The ROS Community

The first dimension of ROS 2 to consider is the *ROS Community*. The Open Source Robotics Foundation[3] greatly enhanced the community of users and developers. ROS 2 is not only a robot programming middleware, but it is also a development methodology, established software delivery mechanisms, and a set of resources made available to members of the ROS community.

ROS 2 is fundamentally *open source*, which means that it is software released under a license in which the user has rights of use, study, change, and redistribution. Many open-source licenses modulate certain freedoms on this software, but essentially we can assume these rights. The most common licenses for ROS 2 software packages

---

[3]https://www.openrobotics.org

are *Apache 2* and *BSD*, although developers are free to use others. I will discuss licenses in greater detail in Section 8.2.1 in the final chapter of this book.

ROS 2 organizes the software following a federal model, providing the technical mechanisms that make it possible. Each developer, company, or institution can develop their software freely, responsible for managing it. It is also widespread that small projects create a community around it, and this community can organize decision-making on releasing issues. These entities create software *packages* for ROS 2 that they can make available in public repositories or be part of a ROS 2 distribution as binaries. Nobody can force these entities to migrate their software to new versions of ROS 2. However, the inertia of many essential and popular packages is enough to guarantee their continuity.

The importance of this development modeling is that it fosters the growth of the ROS community. From a practical point of view, this is key to the success of a robot programming middleware. One of the desirable characteristics of this type of middleware is its support for many sensors and actuators. Nowadays, many manufacturers of these components officially support their drivers for ROS 2 since they know that there are many potential customers and that there are many developers who check if a specific component is supported in ROS 2 before buying them. In addition, these companies usually develop this software in open repositories where user communities can be created reporting bugs and even sending their patches. If you want your library or tool for robots to be widely used, supporting ROS 2 may be the way.

The packages in ROS 2 are organized in distributions. A ROS 2 distribution is made up of a multitude of packages that can work well together. Usually, this implies that it is tied to a specific version of a base system. ROS 2 uses Ubuntu Linux versions as reference. This guarantees stability since otherwise, the dependencies of versions of different packages and libraries would make ROS 2 a real mess. When an entity releases specific software, it does so for a given distribution. It is common to maintain multiple development branches for each distribution.

ROS 2 has released a total of ten distributions to date (November'24), which we can see in Figure 1.2. Each distribution has a name whose initial increases and a different logo (and a different T-shirt model!). An eleventh distribution, which is a bit special, called Rolling Ridley, serves as a staging area for future stable distributions of ROS 2 and as a collection of the most recent development releases.

If you want to contribute your software to a distribution, you should visit the rosdistro repository (`https://github.com/ros/rosdistro`) and a couple of useful links:

- Contributing: `https://github.com/ros/rosdistro/blob/master/CONTRIBUTING.md`

- Releasing your package: `https://docs.ros.org/en/rolling/How-To-Guides/Releasing/Releasing-a-Package.html`

The Open Source Robotics Foundation makes many resources available to the community, among which we highlight:

Figure 1.2: ROS 2 distributions delivered until December 2024.

Distro Name	Release Data	EOL Date	Ubuntu Version
Jazzy Jalisco	May 23rd, 2024	May 2029	Ubuntu 24.04
Iron Irwini	May 23rd, 2023	November 2022	Ubuntu 22.04
Humble Hawksbill	May 23rd, 2022	May 2027	
Galactic Geochelone	May 23rd, 2021	November 2022	Ubuntu 20.04
Foxy Fitzroy	June 5th, 2020	May 2023 (LTS)	
Eloquent Elusor	November 22nd, 2019	November 2020	Ubuntu 18.04
Dashing Diademata	May 31st, 2019	May 2021 (LTS)	
Crystal Clemmys	December 14th, 2018	December 2019	
Bouncy Bolson	July 2nd, 2018	July 2019	Ubuntu 16.04
Ardent Apalone	December 8th, 2017	December 2018	

- ROS Official Page. `http://ros.org`

- ROS 2 Documentation Page: `https://docs.ros.org`. Each distro has its documentation. For example, at `https://docs.ros.org/en/jazzy` you can find installation guides, tutorials, and guides, among others.

- Robotics Stack Exchange (`https://robotics.stackexchange.com`). A place to ask questions and problems with ROS.

- ROS Discourse (`https://discourse.ros.org`). It is a discussion forum for the ROS community, where you can keep up to date with the community, view release announcements, or discuss design issues. They also have ROS 2 user groups in multiple languages.

1.1.2 The Computation Graph

In this second dimension, we will analyze what a robot's software looks like during its execution. This vision will give us an idea of the goal, and we will be able to understand better the why of many of the contents that will follow. This dimension is what we call *Computation Graph*.

A Computation Graph contains ROS 2 nodes that communicate with each other so that the robot can carry out some tasks. The logic of the application is in the nodes, as the primary elements of execution in ROS 2.

ROS 2 makes intensive use of *Object-Oriented Programming*. A node is an *object* of class `Node`, in general, whether it is written in C++ or Python.

A node can access the Computation Graph and provides mechanisms to communicate with other nodes through three types of paradigms:

- **Publication/Subscription**: It is an asynchronous communication where N nodes publish messages to a topic that reaches its M subscribers. A topic is like a communication channel that accepts messages of a unique type. This type of communication is the most common in ROS 2. A very representative case is the node that contains the driver of a camera that publishes images to a topic. All the nodes in a system needing images from the camera to carry out their function subscribe to this topic.

- **Services**: It is an asynchronous communication[4] in which a node requests another node and expects an immediate response. This communication usually requires an immediate response so as not to affect the control cycle of the node that calls the service. An example could be the request to the mapping service to reset a map, with a response indicating if the call succeeded.

- **Actions**: These are asynchronous communications in which a node makes a request to another node. These requests usually take time to complete, and the calling node may periodically receive feedback or the notification that it has finished successfully or with some error. A navigation request is an example of this

---

[4]This communication type was synchronous in ROS 1, but in ROS 2 it is not recommended to implement a synchronous service client (`https://docs.ros.org/en/jazzy/How-To-Guides/Sync-Vs-Async.html`).

type of communication. This request is possibly time-consuming, whereby the node requesting the robot to navigate should not be blocked while completing.

The function of a node in a computational graph is to perform processing or control. Therefore, they are considered active elements with some alternatives in terms of their execution model:

- **Iterative execution**: It is popular in the control software for a node to execute its control cycle at a specific frequency. This approach allows controlling how many computational resources a node requires, and the output flow remains constant. For example, a node calculating motion commands to actuators at 20 Hz based on their status.

- **Event-oriented execution**: The execution of these nodes is determined by the frequency at which certain events occur, usually the arrival of messages at this node. For example, a node that, for each image it receives, performs detection on it and produces an output. The frequency at which an output occurs depends on the frequency at which images arrive. If no images reach it, it produces no output.

Next, we will show several examples of computational graphs. The legend in Figure 1.3 shows the elements used in these examples.

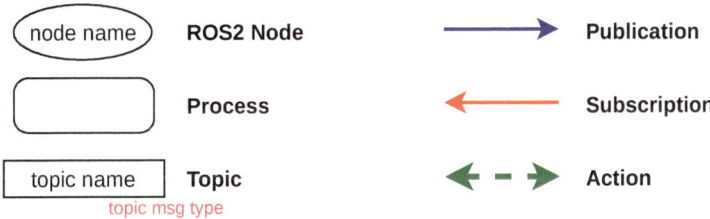

Figure 1.3: Description of symbols used in computer graph diagrams.

The first Computation Graph, shown in Figure 1.4, is a simple example of a program that interacts with a Kobuki[5] robot, a small mobile robot based on Roomba[6].

The Kobuki robot driver is a node that communicates with the robot's hardware using a native driver. Its functionality is exposed to the user through various topics. In this case, we have shown only two topics:

- /mobile_base/event/bumper: A topic in which the Kobuki driver publishes a kobuki_msgs/msg/BumperEvent message every time one of the bumpers changes state (whether or not it is pressed). All nodes of the system interested in detecting a collision with this sensor subscribe to this topic.

---

[5]http://kobuki.yujinrobot.com/about2
[6]https://en.wikipedia.org/wiki/Roomba

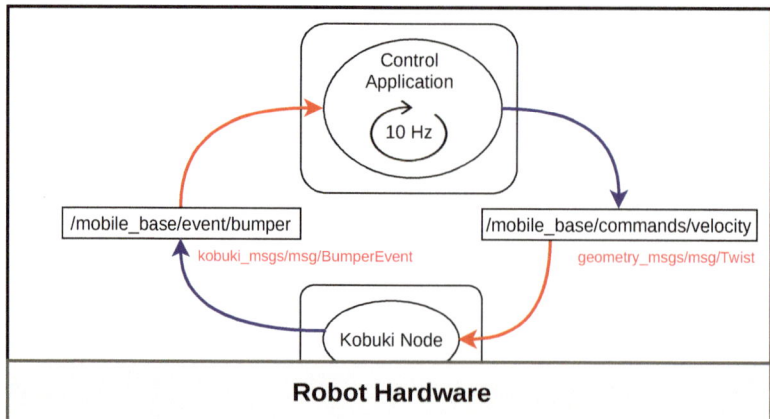

Figure 1.4: Computing graph of a simple control for the Kobuki robot. The control application publishes speeds computed from the information of the bumper to which it subscribes.

- `/mobile_base/commands/velocity`: The topic to which the Kobuki driver subscribes to adjust its speed. If it does not receive any command in a second, it stops. This topic is of type `geometry_msgs/msg/Twist`. Virtually all mobile robots in ROS 2 receive these types of messages to control their speed.

Deep dive: Names in ROS 2

The names of the resources in ROS 2 follow a convention very similar to the filesystem in Unix. When creating a resource, for example, a publisher, we can specify its name as relative, absolute (begins with "/"), or private (begins with "~"). Furthermore, we can define a namespace whose objective is to isolate resources from other namespaces by adding the workspace's name as the first component of the name. Namespaces are helpful, for example, in multirobot applications. Let's see an example of the resulting name of a topic based on the node name and the namespace:

name	Result: (node: my_node / ns: none)	Result: (node: my_node / ns: my_ns)
my_topic	/my_topic	/my_ns/my_topic
/my_topic	/my_topic	/my_topic
~my_topic	/my_node/my_topic	/my_ns/my_node/my_topic

Further readings:

- `http://wiki.ros.org/Names`
- `https://design.ros2.org/articles/topic\_and\_service\_names.html`

This node runs inside a separate process. The Computation Graph shows another process that subscribes to the bumper's topic, and based on the information it receives, it publishes the speed at which the robot should move. We have set the node's execution frequency to indicate that it makes a control decision at 10 Hz, whether or not it receives messages about the status of the bumper.

This Computation Graph comprises two nodes and two topics, with their respective publication/subscription connections.

Let's evolve the robot and the application. Let's add a laser and a 3D camera (also called RGBD camera). For each sensor, a node must access the sensor and present it with a ROS 2 interface. As we said earlier, publishing the data from a sensor is the most convenient way to make this data available in a computational graph.

The application now makes the robot move toward people or objects detected from the 3D image of an RGBD camera. A laser sensor avoids colliding as we move. The Computation Graph shown in Figure 1.5 summarizes the application:

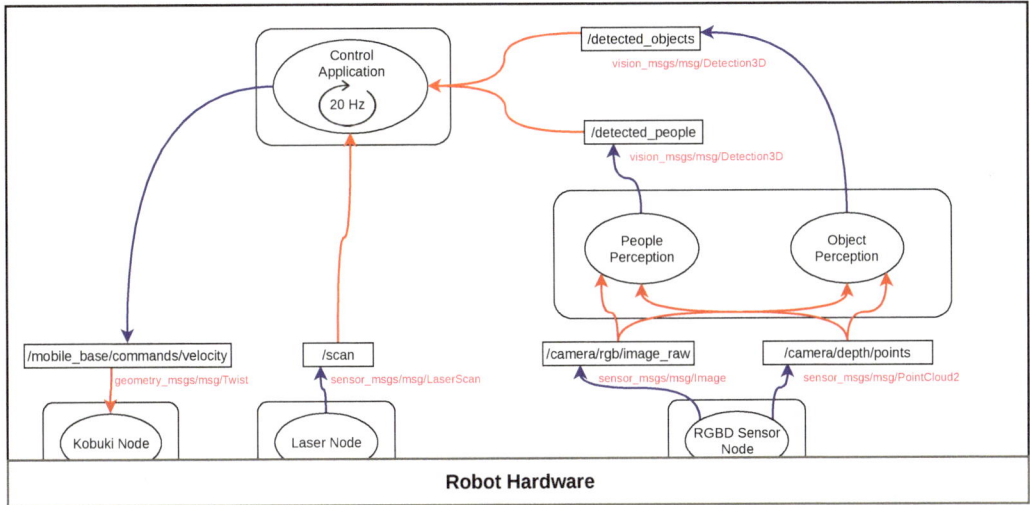

Figure 1.5: Computing graph of a control application that uses the laser data and preprocessed information (people and objects) obtained from the robot's RGBD camera.

- The control node runs at 20 Hz sending control commands to the robot base. It subscribes to the topic `/scan` to check the obstacles around it.

- The process contains two nodes that detect people and objects, respectively. Both need the image and depth information from the camera to determine the position of detected objects. Each detection is published to two different topics, using a standard message designed for 3D detection.

- The control node subscribes to these topics to carry out its task.

Using the Tiago⁷ robot from the previous example, let's assume that there is only one node that provides its functionality. We use in this example two subscribers (speed commands to move its base and trajectory commands to move its neck) and two publishers (laser information and the 3D image from an RGBD camera).

The application (Figure 1.6) is divided into two subsystems, each one in a different process that contains the nodes of each subsystem (we have omitted the details of the topics of each subsystem):

- **Behavior subsystem**: It comprises two nodes that collaborate to generate the robot's behavior. There is behavior coordinator (`Coordinator`) and a node

---

⁷https://pal-robotics.com/es/robots/tiago

that implements an active vision system (`HeadController`). `Coordinator` determines where to look and which points the robot should visit on a map.

- **Navigation subsystem**: This example of a navigation subsystem consists of several nodes. The navigation manager coordinates the planner (in charge of creating routes from the robot's position to the destination) and the controller (which makes the robot follow the created route). The planner needs the map provided by a node that loads the environment map and the robot's position that calculates a location node.

- Communication between both subsystems is done using ROS 2 actions. The Navigation Behavior sets a goal and is notified when it is complete. It also periodically receives progress toward the destination. Actions are also used to coordinate the planner and controller within the navigation system.

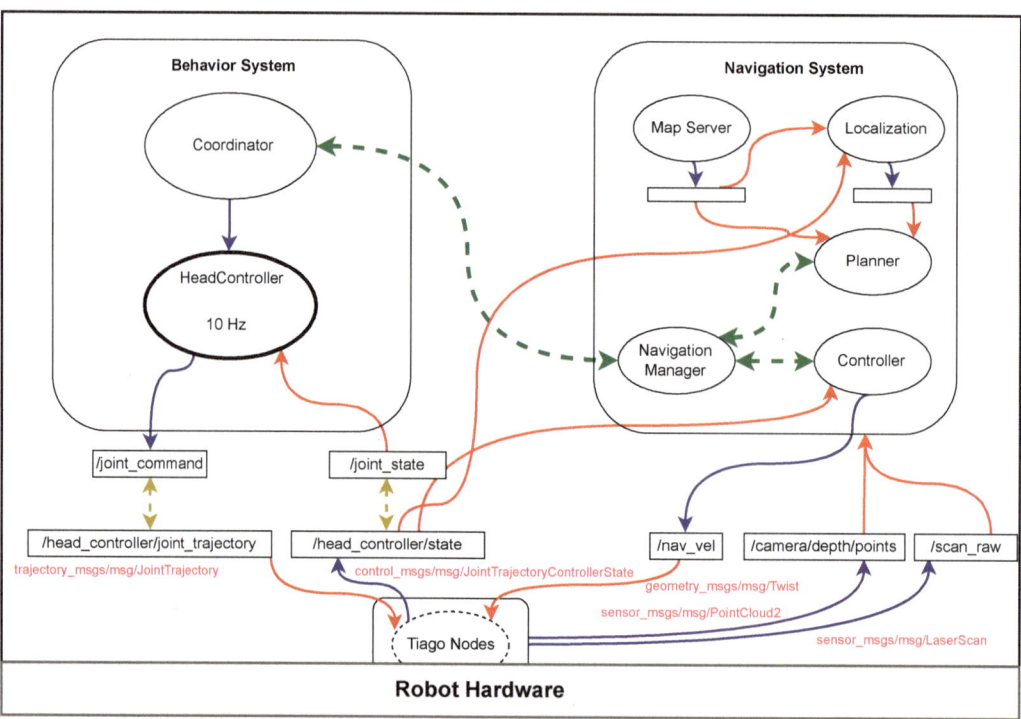

Figure 1.6: Computing graph of behavior-based application for the Tiago robot that uses a navigation subsystem.

Throughout this subsection, we have shown a few examples of computational graphs. Every time we implement an application in ROS 2, we design a computational graph. We establish which nodes we need and what their interactions are. We must decide if a node is executed at a specific frequency or if some event causes its execution (request or message). We can develop all the nodes or include in the Computation Graph nodes developed by third parties.

Although we can define new message types in our application, ROS 2 has defined a set of standard message types that facilitate the interaction of nodes from different developers. It does not make sense, for example, to define a new type of message for images, since there are a lot of third-party software, processing, and monitoring tools that consume and produce the type of message considered standard for images (`sensor_msgs/msg/Image`). Always use existing standard messages whenever possible.

1.1.3 The Workspace

The *Workspace dimension* describes the ROS 2 software from a static point of view. It refers to where the ROS 2 software is installed, organized, and all the tools and processes that allow us to launch a computing graph. This includes the build system and node startup tools.

The fundamental element in this dimension is the *package*. A package contains executables, libraries, or message definitions with a common purpose. Usually, a package *depends* on other packages to run or be built.

Another element of this dimension is the *workspace* itself. A workspace is a directory that contains packages. This workspace has to be activated so that what it contains is available to use.

There can be several workspaces active at the same time. This activation process is accumulative. We can activate an initial workspace that we call *underlay*. Later, we can activate another workspace that we will call *overlay* because it overlays the previous underlay workspace. The overlay package dependencies should be satisfied in the underlay. If a package in the overlay already exists in the underlay, the overlay package hides the one in the underlay.

Usually, the workspace containing the basic ROS 2 installation is activated initially. This is the most common *underlay* in a ROS 2 system. Then, the workspace, where the user is developing their own packages, is activated.

Packages can be installed from sources or with the system installation system. On Ubuntu Linux 24.04, which is the reference in this book, it is carried out with deb packages using tools like *apt*. Each ROS 2 package is packaged in a deb package. The names of deb packages in a distribution are easily identifiable because their names start with `ros-<distro>-<ros2 package name>`. In order to access these packages, configure the APT ROS 2 repository:

```
$ sudo apt update && sudo apt install curl gnupg2 lsb-release

$ sudo curl -sSL https://raw.githubusercontent.com/ros/rosdistro/master/ros.key
-o /usr/share/keyrings/ros-archive-keyring.gpg

$ echo "deb [arch=$(dpkg --print-architecture)
signed-by=/usr/share/keyrings/ros-archive-keyring.gpg]
http://packages.ros.org/ros2/ubuntu $(source /etc/os-release && echo
$UBUNTU_CODENAME) main" | sudo tee /etc/apt/sources.list.d/ros2.list > /dev/null


$ sudo apt-get update
```

Of course, the installation dependencies of the Deb packages are those of the ROS 2 package. The following command shows the ROS 2 packages available to install:

```
$ apt-cache search ros-jazzy
```

ROS 2 Jazzy installation. Instructions are located at https://docs.ros.org/en/jazzy/Installation/Ubuntu-Install-Debs.html. Basically, ROS 2 Jazzy is installed just typing:

```
$ sudo apt update

$ sudo apt install ros-jazzy-desktop
```

All the ROS 2 software installed by apt is in /opt/ros/jazzy. On an Ubuntu 24.04 system, installing the ROS 2 Rolling version is also possible. If it is installed, it is in /opt/ros/rolling. We could even install one of these ROS distributions by compiling its source code in some other location. Because of this, and because it is not recommended (unless you know what you are doing) to mix ROS distributions, installing a distribution does not *activate* it. The activation is done by executing in a terminal:

```
$ source /opt/ros/jazzy/setup.bash
```

This command activates the software in /opt/ros/jazzy. It is common to include this line in $HOME/.bashrc so that it is activated by default when opening a terminal:

```
$ echo "source /opt/ros/jazzy/setup.bash" >> ~/.bashrc
```

It is also convenient to install and configure the *rosdep*[8] tool. This tool locates dependencies not satisfied in a set of source packages and installs them as deb packages. We only need to run these commands once after installation:

```
$ sudo rosdep init

$ rosdep update
```

Typically, the user creates a directory in his $HOME directory that contains the sources of the packages he is developing. Let's create a workspace only by creating a directory with an `src` directory within. Then, add the example packages that we will use throughout this book.

```
$ cd

$ mkdir -p bookros2_ws/src

$ cd bookros2_ws/src

$ git clone -b jazzy-devel https://github.com/fmrico/book_ros2.git
```

---

[8]http://wiki.ros.org/rosdep

If we explore the content that we have added under `src`, we will be able to see a collection of directories. Packages are those that have a `package.xml` file in their root directory.

In this workspace, there are many packages with dependencies on other packages not part of the ROS 2 distribution. To add the sources of these packages to the workspace, we will use the *vcstool*[9]:

```
$ vcs import . < book_ros2/third_parties.repos
```

The command `vcs` reads a list of repositories from a `.repos` file and clones them into the specified directory. Before building, let's use `rosdep` to install any package missing to build the entire workspace:

```
$ rosdep install --from-paths src --ignore-src -r -y
```

Once the sources of the example packages with their dependencies are in the working workspace, build the workspace, always from its root, using the `colcon`[10] command:

```
$ colcon build --symlink-install
```

Check that three directories have been created in the workspace root:

- **build**: Contains the intermediate files of the compilation, as well as the tests, and temporary files.

- **install**: Contains the compilation results, along with all the files necessary to execute them (specific configuration files, node startup scripts, maps ...). Building the workspace using the `--symlink-install` option, creates a symlink to their original locations (in `src` or `build`), instead of copying. This way, we save space and can modify certain configuration files directly in `src`.

- **log**: Contains a log of the compilation or testing process.

Deep dive: `colcon`

`colcon` (collective construction) is a command line tool for building, testing, and using multiple software packages. With colcon, you can compile ROS 1, ROS 2, and even plain CMake packages. It automates the process of building and set up the environment to use the packages.

Further readings:

- `https://design.ros2.org/articles/build\_tool.html`
- `https://colcon.readthedocs.io`
- `https://vimeopro.com/osrfoundation/roscon-2019/video/379127725`

To clean/reset a workspace, simply delete these three directories. A new compilation will regenerate them.

---

[9]`https://github.com/dirk-thomas/vcstool`
[10]`https://colcon.readthedocs.io/en/released/index.html`

In order to use the content of the workspace, activate it as an overlay, in a similar way to how the underlay was activated:

```
$ source ~/bookros2_ws/install/setup.bash
```

It is common to include this line, as well as the underlay, in $HOME/.bashrc so that it is activated by default when opening a terminal:

```
$ echo "source ~/bookros2_ws/install/setup.bash" >> ~/.bashrc
```

1.2 THE ROS 2 DESIGN

Figure 1.7 shows the layers that compose the design of ROS 2. The layer immediately below the nodes developed by the users provides the programmer with an API to interact with ROS 2. Packages in which nodes and programs are implemented in C++ use the C++ client libraries, *rclcpp*. Packages in Python use *rclpy*.

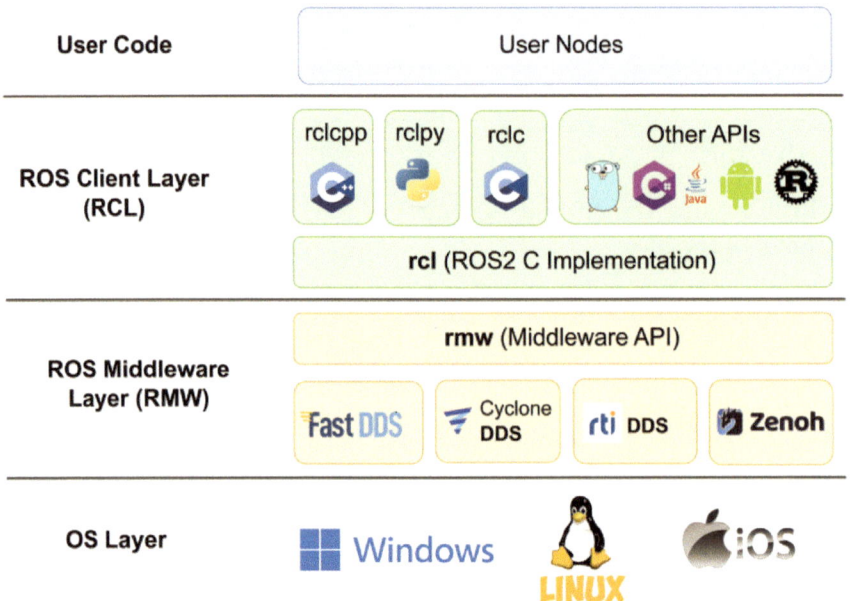

Figure 1.7: ROS 2 layered design.

Rclcpp and rclpy are not completely independent ROS 2 implementations. If so, a node in Python could have different behavior than one written in C++. Both rclcpp and rclpy use *rcl*, which implements the basic functionality of all ROS 2 elements. Rclcpp and rclpy adapt this functionality to the particularites of each language, along with other changes required at that level, such as the threads model.

Any client library for another language (Rust[11], Go[12], Java[13], .NET[14], ...) should be built on top of rcl.

Rcl is the core of ROS 2. No one uses it directly for their programs. There is a C client library for ROS 2 called *rclc* if the user wants to develop nodes written in C. Although it is written in the same language as rcl, it still has to complete some functionality and makes ROS 2 programming less arid than programming using rcl directly.

A crucial component of ROS 2 is communications. ROS 2 is a distributed system whose computing graph has nodes that can be spread over several machines. Even with all the software running on a single robot, nodes are running on the operator's PC to monitor and control the robot's operation.

ROS 2 has initially chosen *Data Distribution Service (DDS)*[15] for its communications layer, a next-generation communications middleware implemented over UDP. It allows the exchange of information between processes with real-time characteristics, security capabilities, and custom quality of service of each connection. DDS provides a publication/subscription communications paradigm, providing a mechanism to discover publishers and subscribers without needing a centralized service automatically. This discovery is made using multicast, although subsequent connections are unicast by default.

There are several DDS vendors, including FastDDS[16], CycloneDDS[17], or RTI[18] Connext. All of them fully or partially implement the DDS standard defined by the OMG[19]. ROS 2 can use all of these DDS implementations. Very few ROS 2 developers would notice using one or the other. However, when we require high performance in latency, amount of data, or resources used, their differences necessitate choosing one that satisfies our criteria.

In recent distributions, initially introduced in Iron and stabilized in Jazzy, the *Zenoh* protocol has emerged as an alternative to DDS for ROS 2 middleware[20]. Zenoh is a high-performance, real-time communication protocol specifically designed to address the challenges of data exchange in distributed systems, particularly in scenarios with stringent requirements for bandwidth, latency, and scalability. Combining features such as publish/subscribe, query/reply, and geo-distributed storage, Zenoh offers a versatile solution for diverse applications. Its integration with ROS 2 is especially significant, as it enhances ROS 2's communication layer by enabling more efficient data dissemination across networks, which is crucial for edge computing scenarios. Zenoh supports various transport protocols (e.g., UDP, TCP, etc.) and can bridge data across geographically distributed locations, effectively addressing some

---

[11]https://github.com/ros2-rust/ros2_rust
[12]https://github.com/tiiuae/rclgo
[13]https://github.com/ros2-java/ros2_java
[14]https://github.com/ros2-dotnet/ros2_dotnet
[15]https://www.omg.org/omg-dds-portal
[16]https://github.com/eProsima/Fast-DDS
[17]https://github.com/eclipse-cyclonedds/cyclonedds
[18]https://www.rti.com/products/dds-standard
[19]https://www.omg.org
[20]https://github.com/ros2/rmw_zenoh

of the limitations of ROS 2's default communication protocol (DDS), particularly in large-scale or geographically dispersed robotic systems.

The APIs of these DDS implementations do not have to be the same. In fact, they are not. For this reason, and to simplify the rcl layer, an underlying layer called *rmw* has been implemented, which presents the rcl programmer with a unified API to access the functionality of each supported middleware implementation. Selecting which DDS to use is trivial, requiring just modifying an environment variable, `RMW_IMPLEMENTATION`.

In the Jazzy distribution, the official DDS implementation is FastDDS. There has been only one exception to this: in the Galactic distribution, CycloneDDS was the official version. This indicates that the official DDS implementation may change again in the future. The competing vendors, in what has humorously been termed the *DDS Wars*, strive to become the leading implementation. Ideally, this competition will ultimately benefit the ROS 2 community.

1.3 ABOUT THIS BOOK

This book is intended to be a journey through programming robots in ROS 2, presenting several projects where the main ROS 2 concepts are applied. Prior knowledge of ROS/ROS 2 is not needed. Many of the concepts presented will sound familiar to ROS 1 programmers along with interesting changes that ROS 2 introduces over the previous version.

We will use C++ as the book's preferred language, although our first examples also include a Python one. We can develop complex and powerful projects in Python, but in my experience with robots, I prefer to use a compiled language rather than an interpreted one. Similarly, the concepts explained with C++ are equally valid with Python. Another decision is to use Linux (specifically Ubuntu GNU / Linux 24.04 LTS) instead of Windows or Mac since it is the reference platform in ROS 2 and the one that I consider predominant in Robotics.

I will assume that the reader is a motivated engineering student or an experienced engineer/professional. We will be using many C++ features up to C++17, including smart pointers (`shared_ptr` and `unique_ptr`), containers (`vector`, `list`, `map`), generic programming, and more. I will try to explain complex code parts from a language point of view, but the less advanced reader may need to consult some references[21,22]. I also count on the reader to know CMake, Git, gdb, and other common tools developers use in Linux environments. It can be a great time to learn it if you do not know it because everything used in this book is what a robot programmer is expected to know.

This book is expected to be read sequentially. It would be difficult for a beginner in ROS 2 to follow the concepts if chapters were skipped. At some points, I will include a text box like this:

---

[21] https://en.cppreference.com/w

[22] https://www.cplusplus.com

> **Deep dive**: Some topic
>
> Some explanation.

This box indicates that in the first reading, it can be skipped and returned to later to deepen our understanding of some concepts.

Throughout the book, I will type shell commands. ROS 2 is used from the shell mainly, and it is important that the user masters these commands. I will use these text boxes for commands in the terminal:

```
$ ls /usr/lib
```

This book is not intended to be a new ROS 2 tutorial. The ones on the official website are great! In fact, there are many concepts (services and actions) that are best learned in these tutorials. This book wants to teach ROS 2 by applying concepts to examples in which a simulated robot performs some mission. Also, we want to teach not only ROS 2 but also some general concepts in Robotics and how they are applied in ROS 2.

Therefore, we will analyze a significant amount of code in this book. I have prepared a repository with all the code that we will use in:

https://github.com/fmrico/book_ros2

At the end of each chapter, I will provide exercises or improvements to help strengthen your understanding of the subject. If you manage to solve it, it can be uploaded to the official repository of the book, in a separate branch with a description and with your authorship. Do this by making a pull request to the official book repository. If I have time (I hope so), I would be happy to review it and discuss it with you.

When it comes to indicating what the structure of a package is, I will use this box:

> Package `my_package`
>
> ```
> my_package/
> ├── CMakeLists.txt
> └── src
>     └── hello_ros.cpp
> ```

To show source code, I will use this other box:

> `src/hello_ros.cpp`
>
> ```cpp
> 1    #include <iostream>
> 2
> 3    int main(int argc, char * argv[]) {
> 4      std::cout << "hello ROS 2" << std::endl;
> 5
> 6      return 0;
> 7    }
> ```

Moreover, when it's just a snippet of code, I will use this kind of box, unnumbered:

```
std::cout << "hello ROS 2" << std::endl;
```

I hope you enjoy this book. Let's start our journey programming robots with ROS 2.

## CHAPTER 2

First Steps with ROS 2

THE previous chapter introduced the fundamental theoretical concepts of ROS 2, in addition to installing ROS 2. In this chapter, we begin to practice with ROS 2 and learn the first ROS 2 concepts.

2.1   FIRST STEPS WITH ROS 2

ROS 2 has been already installed, and both the underlay (`/opt/ros/jazzy`) and the overlay (`~/bookros2_ws`) have been activated, by adding the source commands to `~/.bashrc`. Check it typing:

```
$ ros2

usage: ros2 [-h] Call 'ros2 <command> -h' for more detailed usage. ...
ros2 is an extensible command-line tool for ROS 2.
...
```

If the underlay is activated, this command will be found.

ros2 is the main command in ROS 2. It allows to interact with the ROS 2 system to obtain information or carry out actions.

```
ros2 <command> <verb> [<params>|<option>]*
```

To obtain the list of available packages, type:

```
$ ros2 pkg list

ackermann_msgs
action_msgs
action_tutorials_cpp
...
```

In this case, **pkg** manages ROS 2 packages. The **list** verb obtains the list of packages in the underlay or any overlay.

> **Deep dive**: *roscli*
>
> **ros2cli** is the ROS 2 command line interface tool. It is modular and extensible, so that more functionality can be added by adding new actions. The standard actions currently are:
>
> | | | | |
> |---|---|---|---|
> | action | extension_points | node | test |
> | bag | extensions | param | topic |
> | component | interface | pkg | wtf |
> | launch | run | daemon | lifecycle |
> | security | doctor | multicast | service |
> | control | | | |
>
> Further readings:
>
> - https://github.com/ros2/ros2cli
> - https://github.com/ubuntu-robotics/ros2_cheats_sheet/blob/master/cli/cli_cheats_sheet.pdf

The **ros2** command supports tab-key autocompletion. Type **ros2** and then hit the tab key twice to see the possible verbs. The arguments of a verb can also be discovered with the tab key.

It is also possible to obtain information on a specific package. For example, to get the executable programs from the demo_nodes_cpp package:

```
$ ros2 pkg executables demo_nodes_cpp

demo_nodes_cpp add_two_ints_client
demo_nodes_cpp add_two_ints_client_async
demo_nodes_cpp add_two_ints_server
demo_nodes_cpp allocator_tutorial
...
demo_nodes_cpp talker
...
```

Execute one of them with the command using the **run** verb, which requires two arguments: the package where the executable is and the name of the executable program: The name of this package indicates that all the programs it contains are written in C++.

```
$ ros2 run demo_nodes_cpp talker

[INFO] [1643218362.316869744] [talker]: Publishing: 'Hello World: 1'
[INFO] [1643218363.316915225] [talker]: Publishing: 'Hello World: 2'
[INFO] [1643218364.316907053] [talker]: Publishing: 'Hello World: 3'
...
```

As can be seen, when specifying the program to be executed with the package name and executable name, it is not necessary to know exactly where the programs are, nor to execute them in any specific location.

If everything went well, "Hello world" messages appear in the terminal with a counter. Keep this command running and open another terminal to see what this executable is doing. It is common in ROS 2 to have several terminals open simultaneously, so it is essential to organize them well on the screen to avoid getting lost.

The small Computation Graph created by running the executable is shown in Figure 2.1.

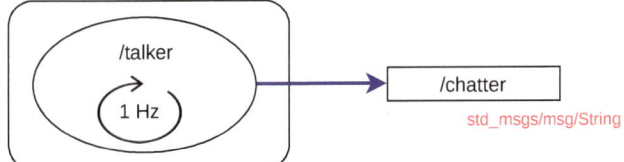

Figure 2.1: Computation Graph for the `Talker` node.

Check the nodes that are currently running using the **node** verb and its **list** argument, executing in another terminal:

```
$ ros2 node list
/talker
```

This command confirms that there is only one node called `/talker`. The names of the resources in ROS 2, as is the case of the nodes, have a similar format to the files in a Linux system. The slash separates parts of the name, starting with the `/` root.

The node `/talker` does not just print an information message through the terminal. It is also publishes messages to a topic.

Check, while the node `/talker` is running, what topics are in the system. For this, use the **topic** verb with its **list** argument.

```
$ ros2 topic list
/chatter
/parameter_events
/rosout
```

There are several topics, including `/chatter`, which is the one that publishes `/talker`. Use the **info** parameter of the **node** verb to get more information:

```
$ ros2 node info /talker
/talker
   Subscribers:
       /parameter_events: rcl_interfaces/msg/ParameterEvent
   Publishers:
       /chatter: std_msgs/msg/String
       /parameter_events: rcl_interfaces/msg/ParameterEvent
       /rosout: rcl_interfaces/msg/Log
   Service Servers:
...
```

The output shows several publishers, which coincide with the topics shown by the previous command since there are no other nodes in the system.

As we have said, each topic supports messages of only one type. The previous commands already showed the type, although it can be verified by asking the `topic` verb directly for the information of a specific topic:

```
$ ros2 topic info /chatter

Type: std_msgs/msg/String
Publisher count: 1
Subscription count: 0
```

Messages are defined in packages that, by convention, have names ending in _msgs. However, there is a growing trend toward naming new message packages with the _interfaces suffix, as these packages often include definitions for services and actions as well. std_msgs/msg/String is the String message defined in the std_msgs package. To check what messages are valid in the system, use the interfaces action and its list argument.

```
$ ros2 interface list

Messages:
    ackermann_msgs/msg/AckermannDrive
    ackermann_msgs/msg/AckermannDriveStamped
    ...
    visualization_msgs/msg/MenuEntry
Services:
    action_msgs/srv/CancelGoal
    ...
    visualization_msgs/srv/GetInteractiveMarkers
Actions:
    action_tutorials_interfaces/action/Fibonacci
    ...
```

The output shows all the types of interfaces through which the nodes can communicate in ROS 2. Adding the -m option, you can filter only the messages. Note that there are more interfaces than just messages. Services and actions also have a format that we can also inspect with `ros2 interface`.

Check the message format to get the fields contained in the message and their type:

```
$ ros2 interface show std_msgs/msg/String

... comments
string data
```

This message format has only one field called data, of type string.

A message is made up of fields. Each field has a different type, which can be a basic type (bool, string, float64) or a message type. In this way, it is usual to create more complex messages from simpler messages.

Stamped messages are an example. A series of messages, whose name ends in **Stamped**, add a header to an existing message. Check the difference between these two messages:

```
geometry_msgs/msg/Point
geometry_msgs/msg/PointStamped
```

Further readings:

- https://docs.ros.org/en/jazzy/Concepts/About-ROS-Interfaces.html

Check the messages currently being published (/talker should be still running in the other terminal) to the topic by typing:

```
$ ros2 topic echo /chatter
data: 'Hello World: 1578'
---
data: 'Hello World: 1579'
...
```

Next, execute a program that contains a node that subscribes to the topic /chatter and displays the messages it receives on the screen. To execute it without stopping the program that contains the /talker node, we run the /listener node, which is in the homonymous program. Although there is a listener node in the demo_nodes_cpp package, for variety, run the listener from a package where the nodes are implemented in Python:

```
$ ros2 run demo_nodes_py listener
[INFO] [1643220136.232617223] [listener]: I heard: [Hello World: 1670]
[INFO] [1643220137.197551366] [listener]: I heard: [Hello World: 1671]
[INFO] [1643220138.198640098] [listener]: I heard: [Hello World: 1672]
...
```

Now the Computation Graph is made up of two nodes that communicate through the topic /chatter. The Computation Graph would look like as shown in Figure 2.2.

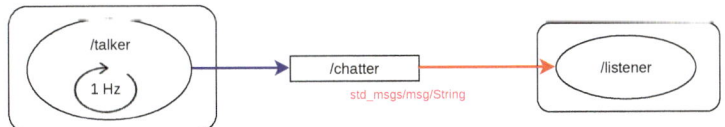

Figure 2.2: Computation Graph for the Listener node.

It is also possible to visualize the Computation Graph by running the rqt_graph tool (Figure 2.3), which is in the rqt_graph package:

```
$ ros2 run rqt_graph rqt_graph
```

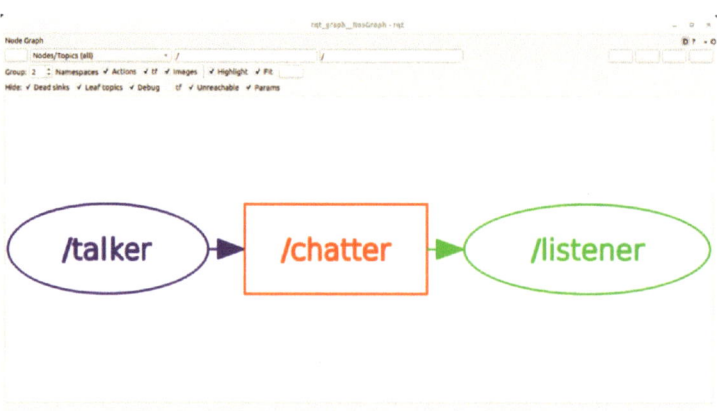

Figure 2.3: Program `rqt_graph`.

Now, stop all the programs by pressing Ctrl-C in the terminals where they are running.

2.2   DEVELOPING THE FIRST NODE

Up to this point, we have only used software from the packages that are part of the ROS 2 base installation. In this section, we will create a package to develop the first node.

The new package will be created in the overlay (`cd ~/bookros2_ws`) to practice creating packages from scratch.

All packages must be in the `src` directory. This time, we use the `ros2` command and the `pkg` verb with the **create** option. In ROS 2 packages, it is necessary to declare what other packages they depend on, either on this workspace or another, so that the compilation tool knows the order they have to be built. Go to the `src` directory and run:

```
$ cd ~/bookros2_ws/src

$ ros2 pkg create my_package --dependencies rclcpp std_msgs
```

This command creates the skeleton of the basics package, with some empty directories to host the source files of our programs and libraries. ROS 2 recognizes that a directory contains a package because it has an XML file called `package.xml`. The **--dependencies** option allows you to add the dependencies of this package. For now, we will use rclcpp, which are the C++ client libraries.

```
package.xml

1   <?xml version="1.0"?>
2   <?xml-model href="http://download.ros.org/schema/package_format3.xsd"
3    schematypens="http://www.w3.org/2001/XMLSchema"?>
4   <package format="3">
5     <name>my_package</name>
6     <version>0.0.0</version>
7     <description>TODO: Package description</description>
8     <maintainer email="john.doe@evilrobot.com">johndoe</maintainer>
9     <license>TODO: License declaration</license>
10
11    <buildtool_depend>ament_cmake</buildtool_depend>
12
13    <depend>rclcpp</depend>
14    <depend>std_msgs</depend>
15
16    <test_depend>ament_lint_auto</test_depend>
17    <test_depend>ament_lint_common</test_depend>
18
19    <export>
20      <build_type>ament_cmake</build_type>
21    </export>
22  </package>
```

Although `ros2 pkg create` is a good starting point for creating a new package, in practice, it is usually made by duplicating an existing package, immediately changing the name of the package, and later adapting it to its purpose.

As the example is a C++ package, since we have indicated that it depends on rclcpp, in its root, a `CMakeLists.txt` file has also been created that establishes the rules to compile it. We will analyze its content as soon as we add something to compile.

First, create the program in ROS 2, as simple as possible, and call it `src/simple.cpp`. The next text boxes contain the package structure and the source code of `src/simple.cpp`:

```
Package my_package

my_package/
├── CMakeLists.txt
├── include
│   └── my_package
├── package.xml
└── src
    └── simple.cpp
```

```
src/simple.cpp

1   #include "rclcpp/rclcpp.hpp"
2
3   int main(int argc, char * argv[]) {
4     rclcpp::init(argc, argv);
5
6     auto node = rclcpp::Node::make_shared("simple_node");
7
8     rclcpp::spin(node);
9
10    rclcpp::shutdown();
11
12    return 0;
13  }
```

- `#include "rclcpp/rclcpp.hpp"` allows access to most of the ROS 2 types and functions in C++.

- `rclcpp::init(argc, argv)` extracts from the arguments with which this process was launched any option that should be taken into account by ROS 2.

- Line 6 creates a ROS 2 node. **node** is a `std::shared_ptr` to a ROS 2 node whose name is `simple_node`.

  The `rclcpp::Node` class is equipped with many aliases and static functions to simplify the code. `SharedPtr` is an alias for `std::shared_ptr<rclcpp::Node>`, and `make_shared` is a static method for `std::make_shared<rclcpp::Node>`.

  The following lines are equivalent, going from a pure C++ statement to one that takes advantage of ROS 2 facilities:

  ```
  1. std::shared_ptr<rclcpp::Node> node = std::shared_ptr<rclcpp::Node>(
     new rclcpp::Node("simple_node"));

  2. std::shared_ptr<rclcpp::Node> node = std::make_shared<rclcpp::Node>(
     "simple_node");

  3. rclcpp::Node::SharedPtr node = std::make_shared<rclcpp::Node>(
     "simple_node");

  4. auto node = std::make_shared<rclcpp::Node>("simple_node");

  5. auto node = rclcpp::Node::make_shared("simple_node");
  ```

- In this code, **spin** blocks the execution of the program so that it does not terminate immediately. This important functionality will be explained in the following examples.

- `shutdown` manages the shutdown of a node prior to the end of the program in the next line.

Examine the `CMakeLists.txt` file, already prepared to compile the program. Some parts that are not relevant have been removed for clarity:

```
CMakeLists.txt
1   cmake_minimum_required(VERSION 3.5)
2   project(basics)
3
4   find_package(ament_cmake REQUIRED)
5   find_package(rclcpp REQUIRED)
6
7   set(dependencies
8     rclcpp
9   )
10
11  add_executable(simple src/simple.cpp)
12  ament_target_dependencies(simple ${dependencies})
13
14  install(TARGETS
15    simple
16    ARCHIVE DESTINATION lib
17    LIBRARY DESTINATION lib
18    RUNTIME DESTINATION lib/${PROJECT_NAME}
19  )
20
21  if(BUILD_TESTING)
22    find_package(ament_lint_auto REQUIRED)
23    ament_lint_auto_find_test_dependencies()
24  endif()
25
26  ament_export_dependencies(${dependencies})
27  ament_package()
```

Identify several parts in this file:

- In the first part, the packages needed are specified with `find_package`. Apart from `ament_cmake`, which is always needed by `colcon`, just `rclcpp` is especified. It is a good habit to create a `dependencies` variable with the packages that this package depends on since we will have to use this list several times.

- For each executable:

    Compile it: Do it with `add_executable`, indicating the name of the result and its sources. Also, use `ament_target_dependencies` to make headers and libraries from other packages accessible for the current target. There are no dependencies with extra libraries, so just using `ament_target_dependencies` is fine.

    Install it: Indicate where to install the program produced, which generally does not vary. A single `install` instruction will be valid for programs and libraries of the package. In general, install everything needed to deploy and run the program. *If it is not installed, it does not exist.*

Compile the workspace:

```
$ cd ~/bookros2_ws

$ colcon build --symlink-install
```

While most ROS commands can be executed from anywhere within the file system, I cannot stress enough that this particular command must always be run from the root of the workspace ~/bookros2_ws. Failing to do so—such as executing it from the ~/bookros2_ws/src directory—will result in the creation of build, install, and `log` directories in the wrong location, cluttering your workspace with unnecessary files. Moreover, such a compilation will have no practical effect, as these directories will not be part of the workspace activation. Consequently, the latest changes from this misplaced compilation will not be available for execution.

Since we have created a new program, we need a reloaded workspace. Open a new terminal and execute:

```
$ ros2 run my_package simple
```

And let's see what happens: absolutely nothing (Figure 2.4).

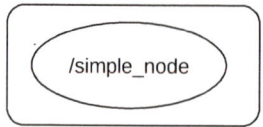

Figure 2.4: Computation Graph for the `Simple` node.

Internally, our program is in the **spin** statement, blocked, waiting for us to finish our program by pressing Ctrl+C. Before doing so, check that the node has been created executing in another terminal:

```
$ ros2 node list
/simple_node
```

We have described how to create a package from scratch. From now on, we will use the packages downloaded from the repository of this book as seen in the previous chapter. This will facilitate more efficient progress without becoming impeded by minor errors during the build of the package, as these can become significant challenges at this stage.

2.3 ANALYZING THE BR2_BASICS PACKAGE

Once this process has been seen in detail, continue analyzing the content of the `br2_basics` package, which contains more interesting nodes. The structure of this package is shown in the following box, and the complete source code can be found in the annexes and in the book repository:

Package `br2_basics`

```
br2_basics
├── CMakeLists.txt
├── config
│   └── params.yaml
├── launch
│   ├── includer_launch.py
│   ├── param_node_v1_launch.py
│   ├── param_node_v2_launch.py
│   ├── pub_sub_v1_launch.py
│   └── pub_sub_v2_launch.py
├── package.xml
└── src
    ├── executors.cpp
    ├── logger_class.cpp
    ├── logger.cpp
    ├── param_reader.cpp
    ├── publisher_class.cpp
    ├── publisher.cpp
    └── subscriber_class.cpp
```

2.3.1 Controlling the Iterative Execution

The previous section described a program containing a node that literally did not do much beyond existing. The program `src/logger.cpp` is more interesting, as it shows more activity:

`src/logger.cpp`

```cpp
auto node = rclcpp::Node::make_shared("logger_node");

rclcpp::Rate loop_rate_period(500ms);
int counter = 0;

while (rclcpp::ok()) {
  RCLCPP_INFO(node->get_logger(), "Hello %d", counter++);

  rclcpp::spin_some(node);
  loop_rate_period.sleep();
}
```

This code shows a common approach to perform a task at a fixed frequency, which is usual in any program that performs some control. The control loop is made in a while loop, controlling the rate with an `rclcpp::Rate` object that makes the control loop stop long enough to adapt to the selected rate.

This code uses **spin_some** instead of **spin**, as used so far. Both are used to manage the messages that arrive at the node, calling the functions that should handle them. While **spin** blocks waiting for new messages, **spin_some** returns once there are no messages left to handle.

As for the rest of the code, **RCLCPP_INFO** is used, which is a macro that prints information. It's very similar to **printf**, passing as first parameter the node's logger (an object inside nodes to log, got with **get_logger** method). These messages are displayed on the screen and are also published to the topic /rosout.

Run this program by typing:

```
$ ros2 run br2_basics logger
[INFO] [1643264508.056814169] [logger_node]: Hello 0
[INFO] [1643264508.556910295] [logger_node]: Hello 1
...
```

The program begins to show messages containing the criticality level of the message, timestamp, the node that produced it, and the message.

As we said before, RCLCPP_INFO also publishes a message of type rcl_interfaces/msg/Log to the topic /rosout, as shown in Figure 2.5. All nodes have a publisher to send the output we generate to this node. It is quite useful when we do not have a console to see these messages.

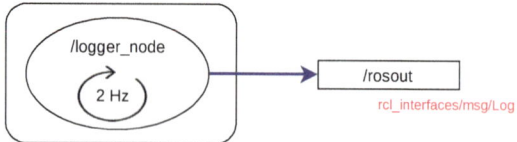

Figure 2.5: Computation Graph for the Logger node.

Take this opportunity to see how to see the messages that are published to a topic:

```
$ ros2 topic echo /rosout
stamp:
    sec: 1643264511
    nanosec: 556908791
level: 20
name: logger_node
msg: Hello 7
file: /home/fmrico/ros/ros2/bookros2_ws/src/book_ros2/br2_basics/src/logger.cpp
function: main
line: 27
---
stamp:
    sec: 1643264512
    nanosec: 57037520
level: 20
...
```

Check the definition of the rcl_interfaces/msg/Log message to verify that the fields shown are the fields of this type of message. In the line field, we have our message:

```
$ ros2 interface show rcl_interfaces/msg/Log
```

Finally, use the rqt_console tool to see the messages that are published in /rosout, as shown in Figure 2.6. This tool, shown in Figure 2.7, is useful when many nodes are generating messages to /rosout, and is useful to filter it by node, by the level of criticality, etc.

```
$ ros2 run rqt_console rqt_console
```

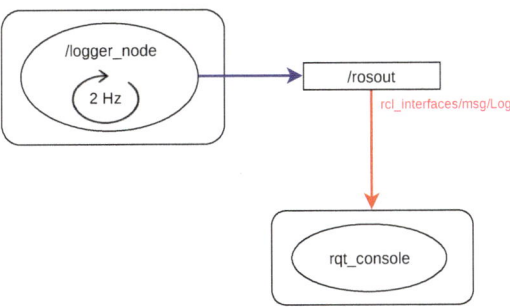

Figure 2.6: `rqt_console` subscribes to `/rosout`, receiving the messages produced by the Logger node.

Figure 2.7: `rqt_console` program.

Test different frequencies by changing the time that object `loop_rate` is created, changing to 100 ms or 1 s, so that the control loop runs at 10 Hz or 1 Hz, respectively.

Do not forget to compile after every change. Use the option `--packages-select` to compile only the package that we have changed, thus saving some time:

```
$ cd ~/bookros2_ws

$ colcon build --symlink-install --packages-select br2_basics
```

From here on, the `cd` command will be omitted. It is essential to perform all the compilations of a workspace from its root directory.

> **Deep dive**: logging
>
> ROS 2 has a logging system that allows generating log messages with increasing severity levels: `DEBUG`, `INFO`, `WARN`, `ERROR` or `FATAL`. For this, use the macro `RCLCPP_[LEVEL]` or `RCLCPP_[LEVEL]_STREAM` to use text streams.
>
> By default, in addition to being sent to `/rosout`, severity levels `INFO` or higher will be displayed on the standard output. You can configure the logger to establish another minimum level of severity to be displayed on the standard output:
>
> ```
> $ ros2 run br2_basics logger --ros-args --log-level debug
> ```
>
> When there are many nodes in an application, it is recommended to use tools such as `rqt_console` that allows selecting nodes and severities.
>
> Further readings:
>
> - https://docs.ros.org/en/jazzy/Tutorials/Logging-and-logger-configuration.html
>
> - https://docs.ros.org/en/jazzy/Concepts/About-Logging.html

The second strategy to iteratively execute a task can be seen in the `src/logger_class.cpp` program. In addition, we show something widespread in ROS 2, which is to implement the nodes inheriting from `rclcpp::Node`. This approach allows for cleaner code and opens the door to many possibilities that will be shown later:

```cpp
src/logger_class.cpp

class LoggerNode : public rclcpp::Node
{
public:
  LoggerNode() : Node("logger_node")
  {
    counter_ = 0;
    timer_ = create_wall_timer(
      500ms, std::bind(&LoggerNode::timer_callback, this));
  }

  void timer_callback()
  {
    RCLCPP_INFO(get_logger(), "Hello %d", counter_++);
  }

private:
  rclcpp::TimerBase::SharedPtr timer_;
  int counter_;
};

int main(int argc, char * argv[]) {
  rclcpp::init(argc, argv);

  auto node = std::make_shared<LoggerNode>();

  rclcpp::spin(node);

  rclcpp::shutdown();
  return 0;
}
```

A timer controls the control loop. This timer produces an event at the desired frequency. When this event happens, it calls the callback that handles it. The advantage is that the node internally adjusts the frequency at which it should be executed without delegating this decision to external code. *Schedule the nodes to know how often they should run.*

To compile these programs, the relevant lines in `CMakeLists.txt` are:

- For each executable, an `add_executable` and its corresponding `ament_target_dependencies`.

- An `install` instruction with all the executables.

```
CMakeLists.txt

1
2   add_executable(logger src/logger.cpp)
3   ament_target_dependencies(logger ${dependencies})
4
5   add_executable(logger_class src/logger_class.cpp)
6   ament_target_dependencies(logger_class ${dependencies})
7
8   install(TARGETS
9     logger
10    logger_class
11    ...
12    ARCHIVE DESTINATION lib
13    LIBRARY DESTINATION lib
14    RUNTIME DESTINATION lib/${PROJECT_NAME}
15  )
```

```
$ ros2 run br2_basics logger_class
```

Build the package and run this program to see that the effect is the same as the previous program. Try to modify the frequencies by setting a different time when creating the timer, in `create_wall_timer`.

2.3.2   Publishing and Subscribing

Now extend the node so that, instead of writing a message on the screen, it publishes a message to a topic (Figure 2.8), publishing consecutive numbers to a topic called `/counter`. An exploration using the `ros2 interface` command with the `list` and `show` options lets you find the message that best suits this duty: `std_msgs/msg/Int32`.

It is necessary to include the headers where it is defined to use a message. Since the type of the message to use is `std_msgs/msg/Int32`, notice how from the name of the message we can easily extract which header to include. Type it while inserting one space before any existing uppercase, and converting all to lowercase. The name of the type is also straightforwards:

```cpp
// For std_msgs/msg/Int32
#include "std_msgs/msg/int32.hpp"

std_msgs::msg::Int32 msg_int32;


// For sensor_msgs/msg/LaserScan
#include "sensor_msgs/msg/laser_scan.hpp"

sensor_msgs::msg::LaserScan msg_laserscan;
```

Now, let's focus on the source code of `PublisherNode`:

```
src/publisher_class.cpp

class PublisherNode : public rclcpp::Node
{
public:
  PublisherNode() : Node("publisher_node")
  {
    publisher_ = create_publisher<std_msgs::msg::Int32>("int_topic", 10);
    timer_ = create_wall_timer(
      500ms, std::bind(&PublisherNode::timer_callback, this));
  }

  void timer_callback()
  {
    message_.data += 1;
    publisher_->publish(message_);
  }

private:
  rclcpp::Publisher<std_msgs::msg::Int32>::SharedPtr publisher_;
  rclcpp::TimerBase::SharedPtr timer_;
  std_msgs::msg::Int32 message_;
};
```

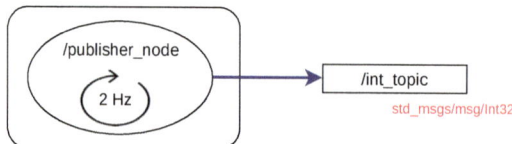

Figure 2.8: Computation Graph for the `Publisher` node.

Let's discuss the important aspects:

- We will use the `std_msgs/msg/Int32` message. From this name, we can deduce that:

  - Its header is `std_msgs/msg/int32.hpp`.
  - The data type is `std_msgs::msg::Int32`.

- Create a publisher, the object in charge of creating the topic (if it does not exist) and publishing the messages. It is possible to obtain more information through this object, such as how many subscribers are listening on a topic. We use `create_publisher`, which is a public method of `rclcpp::Node`, and it returns a `shared_ptr` to an `rclcpp::Publisher` object. The arguments are the name of the topic and an `rclcpp::QoS` object. This class has a constructor that receives an integer that is the size of the output message queue for that topic so that we can put this size directly, and the C++ compiler will do its magic. We will see later that here we can select different QoS.

- We create a `std_msgs::msg::Int32` message, which we can verify that it only has one data field. Every 500 ms, in the timer callback, we increment the message field and call the publisher's `publish` method to publish the message.

Deep dive: QoS in ROS 2

The QoS in ROS 2 is an essential and valuable feature in ROS 2 and a point of failure, so it must be well understood. In the references at the bottom of this table, you can see what QoS policies can be established and their meaning. The following is an example of how to set QoS policies in C++:

```
publisher = node->create_publisher<std_msgs::msg::String>(
  "chatter", rclcpp::QoS(100).transient_local().best_effort());
```

Default	Reliable	Volatile	Keep Last
Services	Reliable	Volatile	Normal Queue
Sensor	Best Effort	Volatile	Small Queue
Parameters	Reliable	Volatile	Large Queue

Each publisher specifies its QoS, and each subscriber can specify its QoS as well. The problem comes because there are QoS that are not compatible, and this will make the subscriber not receive messages:

Compatibility of QoS **Durability** Profiles		**Subscriber**	
		Volatile	**Transient Local**
Publisher	**Volatile**	Volatile	No Connection
	Transient Local	Volatile	Transient Local

Compatibility of QoS **Reliability** Profiles		**Subscriber**	
		Best Effort	**Reliable**
Publisher	**Best Effort**	Best Effort	No Connection
	Reliable	Best Effort	Reliable

The criteria really should be that the publisher should have a less-restrictive QoS policy than the subscriber. For example, the driver of a sensor should publish its readings with a reliable QoS policy. The subscribers decide if they want the communication to be effectively reliable or prefer Best Effort. In this case, these publishers could be:

```
publisher_ = create_publisher<sensor_msgs::msg::LaserScan>(
  "scan", rclcpp::SensorDataQoS().reliable());
```

and the subscribers could use the same QoS, or remove the reliable part.
Further readings:

- https://docs.ros.org/en/jazzy/Concepts/About-Quality-of-Service-Settings.html

- https://design.ros2.org/articles/qos.html

- https://discourse.ros.org/t/about-qos-of-images/18744/16

Run the program:

```
$ ros2 run br2_basics publisher_class
```

And see what we are publishing to the topic:

```
$ ros2 topic echo /int_topic
data: 16
---
data: 17
---
data: 18
...
```

We should see messages with `std_msgs/msg/Int32` messages whose data field is increasing.

Now implement the Node that subscribes to this message:

```
src/subscriber_class.cpp

    class SubscriberNode : public rclcpp::Node
    {
    public:
      SubscriberNode() : Node("subscriber_node")
      {
        subscriber_ = create_subscription<std_msgs::msg::Int32>("int_topic", 10,
          std::bind(&SubscriberNode::callback, this, _1));
      }

      void callback(const std_msgs::msg::Int32::SharedPtr msg)
      {
        RCLCPP_INFO(get_logger(), "Hello %d", msg->data);
      }

    private:
      rclcpp::Subscription<std_msgs::msg::Int32>::SharedPtr subscriber_;
    };
```

In this code, we have created an `rclcpp::Subscription` to the same topic, with the same type of messages. When creating it, we have indicated that for each message published to this topic, the callback function is called, which receives the message in its `msg` parameter as a `shared_ptr`.

Add this program to `CMakeLists.txt`, build, and run `publisher_class` in one terminal and this program in another, composing the Computation Graph shown in Figure 2.9. We will see how the messages received to the topic are displayed on the screen.

```
$ ros2 run br2_basics subscriber_class
```

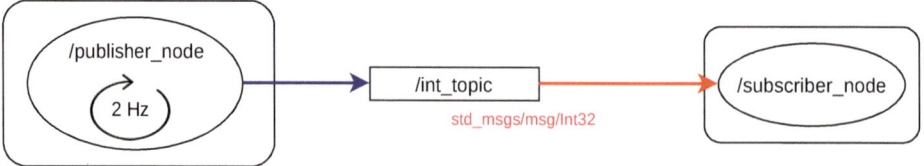

Figure 2.9: Computation Graph for the `Publisher` and `Subscriber` nodes.

2.3.3 Launchers

Up to this point, we have seen that to run a program, we used `ros2 run`. In ROS 2, there is also another way to run programs, which is through the command `ros2 launch` and using a file called launcher, that specifies which programs should be run.

The launcher files are written in Python[1], and their function is declaring which programs to execute with which options or arguments. A launcher can, in turn, include another launcher, allowing you to reuse existing ones.

The need for launchers comes from the fact that a robotic application has many nodes, and they should all be launched simultaneously. Launching one by one and adjusting specific parameters to each one so that the nodes cooperate can be tedious.

Launchers for a package are located in the `launch` directory of a package, and their name usually ends in `_launch.py` or `.launch.py`. Just as `ros2 run` completed with the programs available in a package, `ros2 launch` does the same with the available launchers.

From an implementation point of view, a launcher is a python program that contains a `generate_launch_description()` function that returns a `LaunchDescription` object. A `LaunchDescription` object contains actions, among which we highlight:

- **Node** action: to run a program.

- **IncludeLaunchDescription** action: to include another launcher.

- **DeclareLaunchArgument** action: to declare launcher parameters.

- **SetEnvironmentVariable** action: to set an environment variable.

See how we can launch the publisher and subscriber at the same time. Analyze the first launcher in the `basics` package:

```
launch/pub_sub_v1_launch.py
1   from launch import LaunchDescription
2   from launch_ros.actions import Node
3
4   def generate_launch_description():
5     pub_cmd = Node(
6       package='br2_basics',
7       executable='publisher',
8       output='screen'
9     )
10
11    sub_cmd = Node(
12      package='br2_basics',
13      executable='subscriber_class',
14      output='screen'
15    )
16
17    ld = LaunchDescription()
18    ld.add_action(pub_cmd)
19    ld.add_action(sub_cmd)
20
21    return ld
```

---

[1]Last ROS 2 distros let us to create launchers written in YAML and XML

There is another implementation alternative of this file in launch/pub_sub_v2_launch.py with the same behavior. Check it to see the differences. To use launchers, we must install the launchers directory:

```
CMakeLists.txt

install(DIRECTORY launch DESTINATION share/${PROJECT_NAME})
```

Build the workspace and launch this file:

```
$ ros2 launch br2_basics pub_sub_v2_launch.py
```

In this section, we have seen very simple launchers with very few options. As we progress, we will see more options increasing in complexity.

2.3.4 Parameters

A node uses the parameters to configure its operation. When your program needs configuration files, use parameters. These parameters can be boolean, integer, float, string, or arrays of any of these types. Parameters are read at run time, usually when a node starts, and their operation depends on these values.

Imagine that a node is in charge of locating a robot using a Particle Filter [6] and requires several parameters, such as a maximum number of particles or the topics from which to receive sensory information. This should not be written in the source code since, if we change the robot or environment, these values may be required to be different.

Look at a node that reads these parameters on startup. Create a param_reader.cpp file in the basics package:

```
src/param_reader.cpp

class LocalizationNode : public rclcpp::Node
{
public:
  LocalizationNode() : Node("localization_node")
  {
    declare_parameter("number_particles", 200);
    declare_parameter("topics", std::vector<std::string>());
    declare_parameter("topic_types", std::vector<std::string>());

    get_parameter("number_particles", num_particles_);
    RCLCPP_INFO_STREAM(get_logger(), "Number of particles: " << num_particles_);

    get_parameter("topics", topics_);
    get_parameter("topic_types", topic_types_);

    if (topics_.size() != topic_types_.size()) {
      RCLCPP_ERROR(get_logger(), "Number of topics (%zu) != number of types (%zu)",
        topics_.size(), topic_types_.size());
    } else {
      RCLCPP_INFO_STREAM(get_logger(), "Number of topics: " << topics_.size());
      for (size_t i = 0; i < topics_.size(); i++) {
        RCLCPP_INFO_STREAM(
          get_logger(),
          "\t" << topics_[i] << "\t - " << topic_types_[i]);
      }
    }
  }

private:
  int num_particles_;
  std::vector<std::string> topics_;
  std::vector<std::string> topic_types_;
};
```

- All parameters of a node must be declared using methods like `declare_parameter`. In the declaration, we specify the parameter name and the default value.

- We obtain its value with functions like `get_parameter`, specifying the name of the parameter and where to store its value.

- There are methods to do this in blocks.

- The parameters can be read at any time. You can also subscribe to modifications to these parameters. However, reading them to the startup makes your code more predictable.

If we run our program without assigning a value to the parameters, we will see how the default values take value:

```
$ ros2 run br2_basics param_reader
```

Stop executing the program, and execute our program assigning value to one of the parameters. We can do this in setting arguments, starting with `--ros-args`, and `-p` for setting a parameter:

```
$ ros2 run br2_basics param_reader --ros-args -p number_particles:=300
```

Now pass in values for the remaining parameters. In this case, the two string arrays:

```
$ ros2 run br2_basics param_reader --ros-args -p number_particles:=300
-p topics:= '[scan, image]' -p topic_types:='[sensor_msgs/msg/LaserScan,
sensor_msgs/msg/Image]'
```

If we want to set the parameter values in a launch, we can do it as follows:

```
launch/param_node_v1_launch.py

from launch import LaunchDescription
from launch_ros.actions import Node

def generate_launch_description():
    param_reader_cmd = Node(
        package='br2_basics',
        executable='param_reader',
        parameters=[{
            'particles': 300,
            'topics': ['scan', 'image'],
            'topic_types': ['sensor_msgs/msg/LaserScan', 'sensor_msgs/msg/Image']
        }],
        output='screen'
    )

    ld = LaunchDescription()
    ld.add_action(param_reader_cmd)

    return ld
```

Although this method may be suitable for assigning values to a few parameters, it is usually convenient to use a file containing the parameters' values with which we want to execute a node. This is the way to have configuration files in ROS 2. The chosen format is YAML. Usually, these configuration files are stored in the config directory of our packages, and it is mandatory to mark them to install in the CMakeLists.txt, as it was done with the launch directory:

```
CMakeLists.txt

install(DIRECTORY launch config DESTINATION share/${PROJECT_NAME})
```

Let's discuss an important point: what would prevent someone from using a different organization in their packages? Why name the configuration directory config instead of setup, or startup instead of launch? And why put the source file in another structure? Why use YAML/parameters and not text files or XML and a custom configuration reader? Why use launchers and not a bash script? And why not an application that launches all the necessary nodes?

Of course, a ROS 2 developer could make other decisions, but there are recognized *best practices*[2]. These best practices have the advantage that when another developer tries to use your code, it is much easier to find and identify the critical elements. My recommendation is to follow these conventions. This way, your code can be used by more people, it will be more maintainable in the long term, and may receive more collaborations. This is critical in a business environment as it will ease knowledge transfer and software maintainability as developers come and go.

---

[2]https://docs.ros.org/en/rolling/The-ROS2-Project/Contributing/Developer-Guide.html

Continue with our example. A file with the parameters with our node could look like this:

```
config/params.yaml

localization_node:
  ros__parameters:
    number_particles: 300
    topics: [scan, image]
    topic_types: [sensor_msgs/msg/LaserScan, sensor_msgs/msg/Image]
```

And execute indicating specifying the location of our file. If we have installed the config directory and compiled it, we can execute:

```
$ ros2 run br2_basics param_reader --ros-args --params-file
install/br2_basics/share/br2_basics/config/params.yaml
```

If we want it to be read in a launcher, we will use:

```
launch/param_node_v1_launch.py

1
2    def generate_launch_description():
3        ...
4        param_reader_cmd = Node(
5            package='br2_basics',
6            executable='param_reader',
7            parameters=[param_file],
8            output='screen'
9        )
```

2.3.5 Executors

As the nodes in ROS 2 are C++ objects, a process can have more than one node. In fact, in many cases, it can be very beneficial to do so since communications can be accelerated by using shared memory strategies when communication is within the same process. Another benefit is that it can simplify the deployment of nodes if they are all in the same program. The drawback is that a failure in one node could cause all nodes of the same process to terminate.

ROS 2 offers you several ways to run multiple nodes in the same process. The most recommended is to make use of the *Executors*. An Executor is an object to which nodes are added to execute them together. See an example:

Single thread executor

```cpp
int main(int argc, char * argv[]) {
  rclcpp::init(argc, argv);

  auto node_pub = std::make_shared<PublisherNode>();
  auto node_sub = std::make_shared<SubscriberNode>();

  rclcpp::executors::SingleThreadedExecutor executor;

  executor.add_node(node_pub);
  executor.add_node(node_sub);

  executor.spin();

  rclcpp::shutdown();
  return 0;
}
```

Multi thread executor

```cpp
  auto node_pub = std::make_shared<PublisherNode>();
  auto node_sub = std::make_shared<SubscriberNode>();

  rclcpp::executors::MultiThreadedExecutor executor(rclcpp::ExecutorOptions(), 8);

  executor.add_node(node_pub);
  executor.add_node(node_sub);

  executor.spin();
}
```

In both codes, we create an executor to which we add the two nodes (Figure 2.10) so that the spin call handles both nodes. The difference between the two is using a single thread for this management, or using eight threads to optimize the processor capabilities.

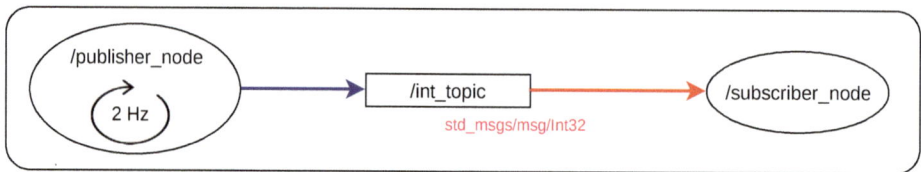

Figure 2.10: Computation Graph for the `Publisher` and `Subscriber` nodes, running in the same process.

2.4   SIMULATED ROBOT SETUP

So far we have seen the basics package, which shows us basic elements of ROS 2, how to create nodes, publications, and subscriptions. ROS 2 is not a communications middleware, but a robot programming middleware, and this book tries to create behaviors for robots. Therefore, we need a robot. Robots are relatively expensive. It is possible to have a real robot, such as the Kobuki (turtlebot 2) equipped with a laser and an RGBD camera for around 1000€. A robot considered professional can go to several tens of thousands of euros. As not all readers have plans to acquire a robot to run ROS 2, we are going to use the Tiago robot in a simulator.

The Tiago robot ("iron" model) from PAL Robotics consist of a differential base with distance sensors and a torso with an arm, an RGBD camera located on its head.

Among the packages that we have already added to the worskspace, there were already those necessary to simulate the Tiago robot in Gazebo (one of the reference simulator in ROS 2). So we will only have to use a launcher that we have created in the br2_tiago package:

```
$ ros2 launch br2_tiago sim.launch.py
```

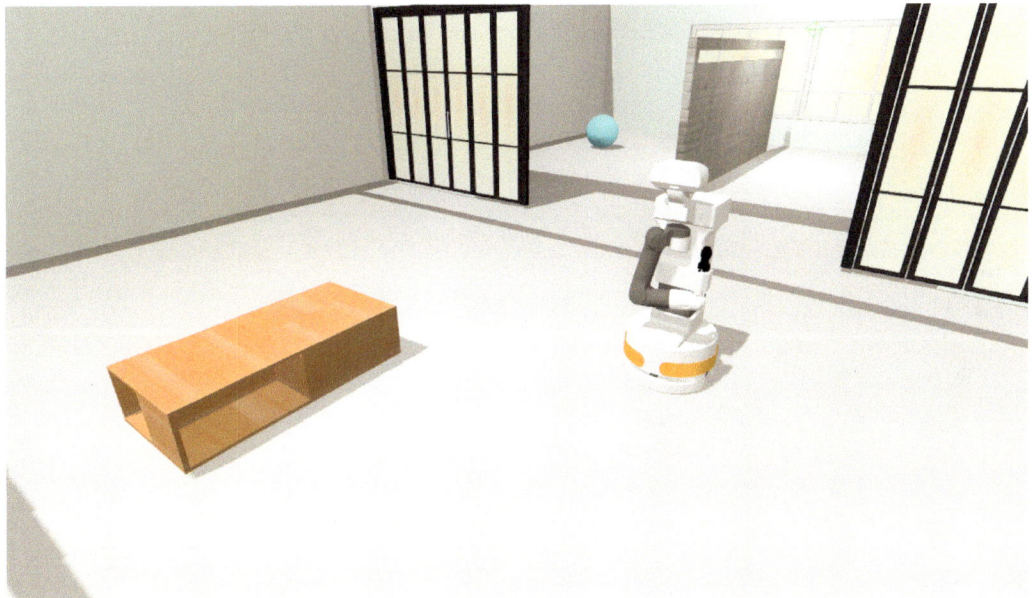

Figure 2.11: Simulating Tiago robot in Gazebo.

There are several worlds available (you can examine ThirdParty/br2_gazebo_worlds). By default, the world that is loaded is home.world. If you want to use a different one, you can use the launcher world parameter, as shown in the following examples:

```
$ ros2 launch br2_tiago sim.launch.py world:=empty
$ ros2 launch br2_tiago sim.launch.py world:=follow_line
```

One of the first things you can do when you use a robot for the first time and have just launched its driver or simulation is to see what topics it provides, either as a publisher or a subscriber. That will be the interface we will use to receive information from the robot and send it commands. Open a new terminal and execute:

```
$ ros2 topic list
```

This will be the main interface with the robot's sensors and actuators. Figure 2.14 shows a non-exhaustive way the nodes and topics that are available to the programmer to interact with the simulated robot:

- Virtually all nodes are within the Gazebo simulator process. Outside there are only two of them:

  /twist_mux Create several subscribers to topics that receive robot speeds, but from different sources (mobile, tablet, keys, navigation, among others), publishing in /cmd_vel_muxed the result of mixing them. The message type of all these topics is textttgeometry_msgs/msg/Twist.

  /twist_stamper Takes the messages of type geometry_msgs/msg/Twist from /cmd_vel_muxed and publish them to /mobile_base_controller/cmd_vel as a geometry_msgs/msg/TwistStamped.

  /robot_state_publisher It is a standard node in ROS 2 that reads the description of a robot from a URDF file and subscribes to the status of each of the robot's joints. In addition to publishing this description in URDF, it creates and updates the robot frames in the TFs system (we will explain the TF system in the next chapters), a system to represent and link the different geometric axes of reference in the robot.

- The nodes on the left take care of the sensors. They publish information from the robot's camera, imu, laser, and sonar. The most complex node is the camera node, an RGBD sensor, since it publishes the depth and RGB images separately. Each type of image has associated a topic **camera_info** that contains the intrinsic values of the robot's camera. For each sensor, the standard message types are used for the information provided.

- The nodes at the bottom use the same interface to move the head, the torso, and the gripper. At the top roght it is also the node which controls the arm. They all use the **joint_trajectory_controller** from the **ros2_control** package.

- The nodes on the right are responsible for the following:

  /joint_state_broadcaster Publish the status of each of the joints of the robot.

  /mobile_base_controller Makes the robot base move with the speed commands it receives. In addition, it publishes the estimated displacement of the base.

First, teleoperate the robot to move it. For this, ROS 2 has several packages that take commands from the keyboard, from a PS or XBox controller, or a mobile phone, and publish geometry_msgs/msg/Twist messages to the topic /cmd_vel. In this case we will use teleop_twist_keyboard. This program receives keystrokes by stdin in publishes /cmd_vel movement commands.

As the topic /cmd_vel of teleop_twist_keyboard does not match any input topic of our robot, we must do a remap. A remap (Figure 2.12) allows you to change the name of one of its topics when executing (at *deployment time*). In this case, we are going to execute teleop_twist_keyboard indicating that instead of publishing in /cmd_vel, publish to the topic /key_vel of the robot:

```
$ ros2 run teleop_twist_keyboard teleop_twist_keyboard --ros-args -r
cmd_vel:=key_vel
```

Now we can use the keys indicated by `teleop_twist_keyboard` to move the robot. Remapping a topic is an important feature of ROS 2 that allows different ROS 2 programs from other developers to work together.

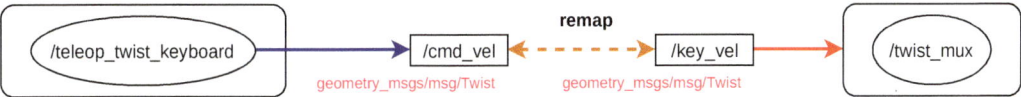

Figure 2.12: Connection between the Tiago and the teleoperator velocity topics, using a remap.

Now is the time to see the robot's sensory information. Until now, we could use `ros2 topic` to see the topics of the camera or the laser with one of these commands:

```
$ ros2 topic echo /scan_raw

$ ros2 topic echo /head_front_camera/image
```

But it is hard to show sensory information, especially if it is so complex. Use the `--no-arr` option so that it does not display the content of the data arrays.

```
$ ros2 topic echo --no-arr /scan_raw

$ ros2 topic echo --no-arr /head_front_camera/image
```

Analyze the information it shows. There is a common field in both messages, which is common in messages with perceptual information and is repeated in many types of messages, especially those that end in the adjective "*Stamped". It has a header of type `std_msgs/msg/Header`. As we have just seen, Messages can be defined by composing basic types (`Int32`, `Float64`, `String`) or already existing messages, like this one.

The header is tremendously helpful for handling sensory information in ROS 2. When a sensor driver publishes messages with its data, it uses the header to tag this reading with:

- The data capture timestamp. Even if a message is received or processed late, the reading can be placed at its corresponding capture moment, supporting some latencies.

- The frame in which it was taken. A frame is an axis of references in which the spatial information (coordinates, distances, etc.) contained in the message makes sense. Usually, each sensor has its frame (even several).

A robot is geometrically modeled using a tree whose tree nodes are the frames of a robot. By convention, a frame should have a single parent frame and all required child frames. The parent–child relationship is through a geometric transformation that includes a translation and a rotation. The frames usually appear at points on the robot subject to variation, as in the case of the motors joining the robot's parts.

ROS 2 has a system called TF, which we will explain in the next chapters, which maintains these relationships through two topics /tf, for geometric transformations that vary, and /tf_static if they are fixed.

ROS 2 has several tools that help us display sensory and geometric information, and perhaps the most popular is RViz2. Start by running it by typing in a terminal:

```
$ ros2 run rviz2 rviz2
```

RViz2 is a viewer that allows to display information contained in the topics. If this is your first time opening RViz2, it will probably appear quite empty; only a grid through which we can navigate using the keys and the mouse. We will discover the information about our robot step by step, as shown in Figure 2.13:

1. On the left, in the Displays panel, RViz2 has some global options in which we have to specify what our Fixed Frame is, that is, the coordinate axis of the 3D visualization shown on the right. For now, we are going to select base_footprint. By convention in ROS 2, this frame is a frame that is in the center of the robot, on the ground, and is a good starting point for our exploration.

2. In the Displays Panel, we are going to add different visualizations. The first will be to see the frames of the robot. Press the Add button, and look in the "By display type" tab, the TF element. All the robot frames will appear instantly. If they seem like a lot to you, display the TF component in the Displays Panel, uncheck the "All Enabled" box and start adding or removing the frames you want.

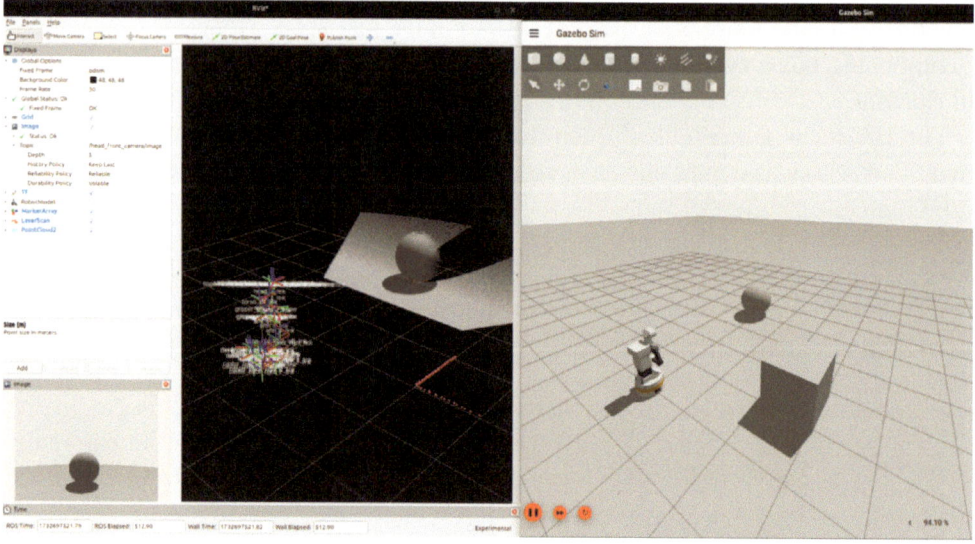

Figure 2.13: RViz2 visualizing the TF tree and the sensor information of the Tiago robot.

3. Add several elements to Gazebo, as seen in Figure 2.13. If not, we will not perceive much either.

4. Add the laser information of the robot. Press Add again and in the "By Topic" tab, select the topic /scan_raw, which already indicates that it is a LaserScan. In the LaserScan element that has been added in the Display Panel, we can see information and change the display options. Display the options for this element:

   - The Status should show ok and have a counter that goes up as it receives messages. If it displays an error, it usually contains information that can help us figure out how to fix it.

   - The Topic has to do with the topic to show and the QoS with which RViz2 subscribes to that topic. If we do not see anything, it may be that we have not selected a compatible QoS.

   - From here on, the rest of the options are specific to this type of message. We can change the size of the dots that represent laser readings, their color, or even the visual element used.

5. As done with the laser, add a visualization of the topic that contains the Point-Cloud2 (/head_front_camera/points).

Use the teleoperator to move the robot. In RViz2, the movement of the robot is not perceived, only the frame odom moving. This is because the center of this visualization is always the Fixed Frame, which we now have as base_footprint. Change the Fixed Frame to odom, which is a Frame that represents the position of the robot when it started. Now we can appreciate the robot's movement around its environment. The odom → base_footprint transform, by convention in ROS 2, collects the translation and rotation calculated by the robot driver from its starting point.

With this, we have explored the capabilities of the simulator robot and various tools for managing our robot.

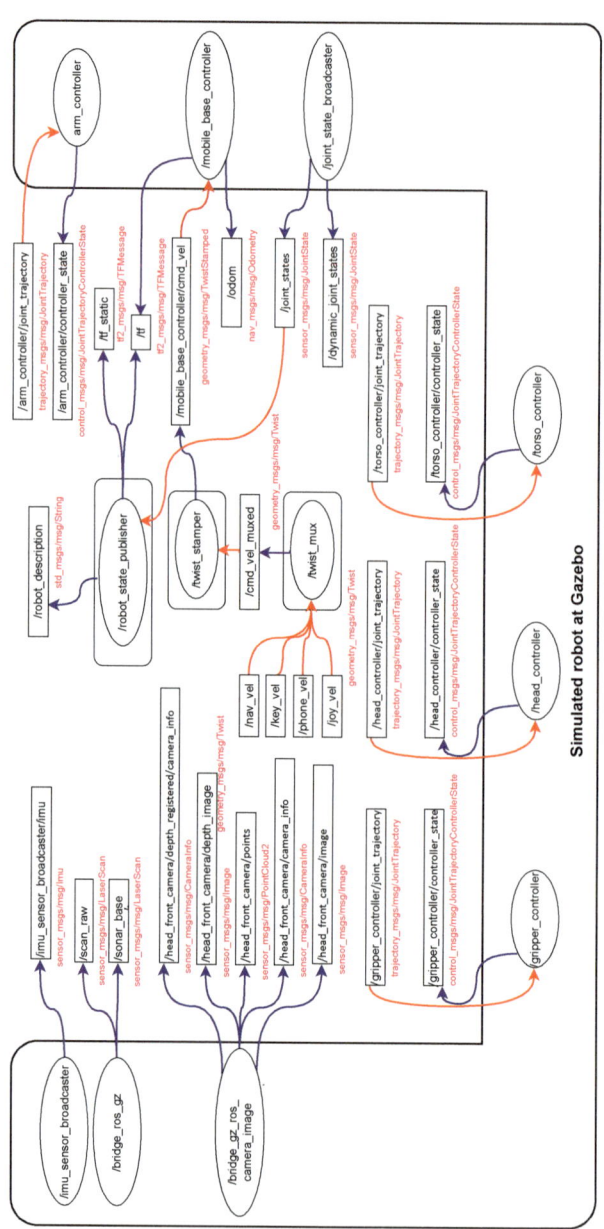

Figure 2.14: Computation Graph for the Tiago robot, displaying the relevant topics.

## CHAPTER 3

First Behavior: Avoiding Obstacles with Finite States Machines

T HIS section aims to apply everything shown until now to create seemingly "smart" behavior. This exercise will put together many things we have presented and show how effective it is to program a robot using ROS 2. In addition, we will address some issues in robot programming.

The *Bump and Go* behavior uses the robot's sensor to detect nearby obstacles in front of the robot. The robot moves forward, and when it detects an obstacle, it goes back and turns for a fixed time to move forward again. Although it is a simple behavior, some decision-making approach is recommended since our code, even if it is simple, can start to fail as we solve subsequent problems that may arise. In this case, we will use a Finite State Machine (FSM) .

An FSM is a mathematical computational model that we can use to define the behavior of a robot. It is made up of states and transitions. A robot keeps producing an output in one state until the condition of an outgoing transition is fulfilled and it transits to the target state of this transition.

Applying an FSM can significantly reduce the complexity of solving a problem when we implement simple behaviors. For a moment, try to think about how to approach the *Bump and Go* problem using loops, ifs, temporary variables, counters, timers. It would be a complex program to understand and follow its logic. Once finished, adding some additional conditions will probably make you throw away what we have done and start over.

Applying an FSM-based solution to the *Bump and Go* problem is straightforward. Think about the different outputs that the robot must produce (stop, move forward, go back, and turn). Each of these actions will have its own state. Now think about the transitions between states (connection and condition), and we will obtain an FSM like the one shown in Figure 3.1.

DOI: 10.1201/9781003516798-3

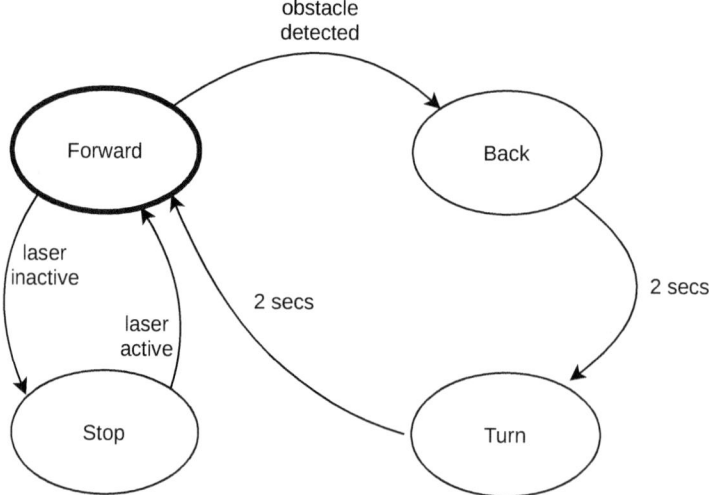

Figure 3.1: States and Transitions for solving the *Bump and Go* problem using an FSM.

3.1 PERCEPTION AND ACTUATION MODELS

This section analyzes what perceptions we use to solve the *Bump and Go* problem and what actions we can produce.

In both models, first of all, we must define the used geometric conventions:

- ROS 2 uses the metric International System of Measurements (SI). For different dimensions, we will consider the units of meters, seconds, and radians. Linear speeds should be m/s, rotational speeds rad/s, linear accelerations m/s^2, and so on.

- In ROS 2, we are right-handed[1] (left part of Figure 3.2): x grows forward, y to the left, and z grows up. If we establish the reference origin on our chest, a coordinate whose x is negative would be behind us, and a positive z would be above us.

- Angles are defined as rotations around the axes. Rotation around x is sometimes called the roll, y pitch, and z yaw.

- Angles grow by rotating counter-clockwise (right part of Figure 3.2). Angle 0 is forward, π is back, and $\pi/2$ is left.

For this problem, we will use the information of the laser sensor, which we saw in the previous chapter that was in the topic /scan_raw, and whose type was sensor_msgs/msg/LaserScan. Check this message format by typing:

---

[1] https://www.ros.org/reps/rep-0103.html

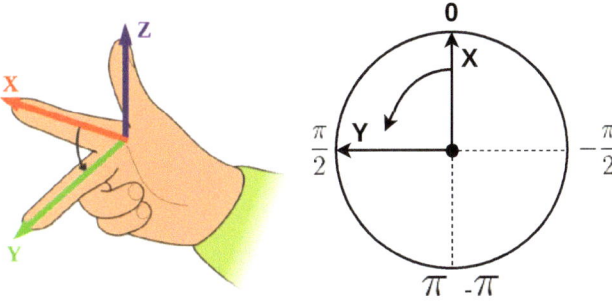

Figure 3.2: Axis and angles conventions in ROS.

```
$ ros2 interface show sensor_msgs/msg/LaserScan

# Single scan from a planar laser range-finder
#
# If you have another ranging device with different behavior (e.g. a sonar
# array), please find or create a different message, since applications
# will make fairly laser-specific assumptions about this data

std_msgs/Header header  # timestamp in the header is the acquisition time of
                        # the first ray in the scan.
                        #
                        # in frame frame_id, angles are measured around
                        # the positive Z axis (counterclockwise, if Z is up)
                        # with zero angle being forward along the x axis

float32 angle_min       # start angle of the scan [rad]
float32 angle_max       # end angle of the scan [rad]
float32 angle_increment # angular distance between measurements [rad]

float32 time_increment  # time between measurements [seconds] - if your scanner
                        # is moving, this will be used in interpolating pos
                        # of 3d points
float32 scan_time       # time between scans [seconds]

float32 range_min       # minimum range value [m]
float32 range_max       # maximum range value [m]

float32[] ranges        # range data [m]
                        # (Note: values < range_min or > range_max should be
                        # discarded)
float32[] intensities   # intensity data [device-specific units]. If your
                        # device does not provide intensities, please leave
                        # the array empty.
```

To see one of these laser messages (without showing the content of the readings), launch the simulator and type:

```
$ ros2 topic echo /scan_raw --no-arr

---
header:
  stamp:
    sec: 11071
    nanosec: 445000000
  frame_id: base_laser_link
angle_min: -1.9198600053787231
angle_max: 1.9198600053787231
angle_increment: 0.005774015095084906
time_increment: 0.0
scan_time: 0.0
range_min: 0.05000000074505806
range_max: 25.0
ranges: '<sequence type: float, length: 666>'
intensities: '<sequence type: float, length: 666>'
---
```

In Figure 3.3, we can see the interpretation of this message. The key is that in the ranges field are the distances to obstacles. Position 0 of this std::vector (arrays in messages are represented as std::vector in C++) corresponds to angle −1.9198, position 1 is this angle plus the increment, until this vector is completed. It is easy to check that if we divide the range (maximum angle minus minimum angle) by the increment, we get these 666 readings, which is the size of the ranges vector.

Most messages, especially if they contain spatially interpretable information, have a header containing the timestamp and the sensor frame. Note that a sensor can be mounted in any position or orientation on the robot, even in some moving parts. The sensor frame must have a geometric connection (a rotation and translation) to the rest. On many occasions, we will need to transform the coordinates of the sensory information to the same frame to fuse it, which is usually base_footprint (the center of the robot, at ground level, pointing forward). These geometric manipulations are explained in the next chapter.

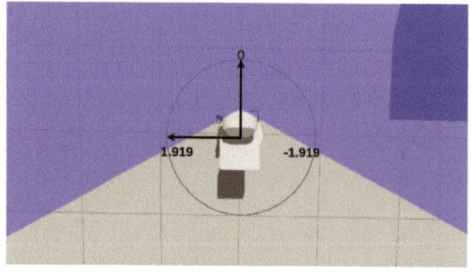 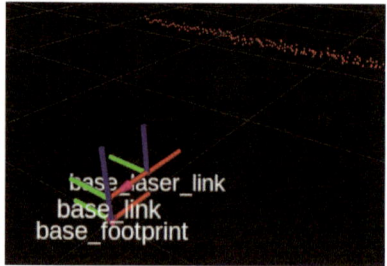

Figure 3.3: Laser scan interpretation in the simulated Tiago (left). Laser frame with respect to other main frames (right).

In our problem, we are only interested in whether there is an obstacle in front of the robot, which is angle 0, and this corresponds exactly to the content of the middle position of the vector of ranges. We can use the original frame of the sensor since it is aligned, a little forward and up, with base_footprint.

An essential feature of ROS is standardization. Once a consensus has been reached in the community on the format in which the information produced by a laser sensor is encoded, all laser driver developers should use this format. This consensus means that the message format must be general enough to support any laser sensor. In the same way, an application developer must exploit the information in this message for his program to function correctly regardless of the characteristics of the sensor that produced the sensory reading. The great advantage of this approach is that we can make any ROS program work with any ROS-supported laser, allowing the software to be truly portable between robots. Also, an experienced ROS developer does not have to learn new, manufacturer-defined formats. Finally, using this format puts at your disposal a wide variety of utilities to filter or monitor laser information. This approach applies to all types of sensors and actuators in ROS, which may be one reason for the success of this framework.

Regarding the **action model** in this example, we will send the robot translation and rotation speeds to topic `/nav_vel`, which is of type `geometry_msgs/msg/Twist`. Let's see this message format:

```
$ ros2 interface show geometry_msgs/msg/Twist

Vector3 linear
Vector3 angular

$ ros2 interface show geometry_msgs/msg/Vector3

float64 x
float64 y
float64 z
```

All robots use this message format to receive speeds, allowing generic teleoperation programs (with keyboards, joysticks, mobiles, etc.) and navigation in ROS. Once again, we are talking about standardization.

The `geometry_msgs/msg/Twist` message is much more generic than what our robot supports. We cannot make it move in Z (it cannot fly) or move laterally with just two wheels since it is a differential robot. We could probably do more translations and rotations if we had a quadcopter. We can only make it go forward or backward, rotate, or combine both. For this reason, we can only use the fields `linear.x` and `angular.z` (rotation to the Z-axis, positive velocities to the left, as indicated in Figure 3.2).

3.2 COMPUTATION GRAPH

The Computation Graph of this application will be pretty simple: A single node that subscribes to the laser topic and publishes speed commands to the robot.

The control logic interprets the input sensory information and produces the control commands. This logic is what we are going to implement with an FSM. The logic control will run iteratively at 20 Hz. The execution frequencies depend on publishing the control commands. If it is not published above 20 Hz, some robots stop, which is very convenient so that there are no robots without control in the laboratory. Commonly, the frequency at which we receive information is not the same as the

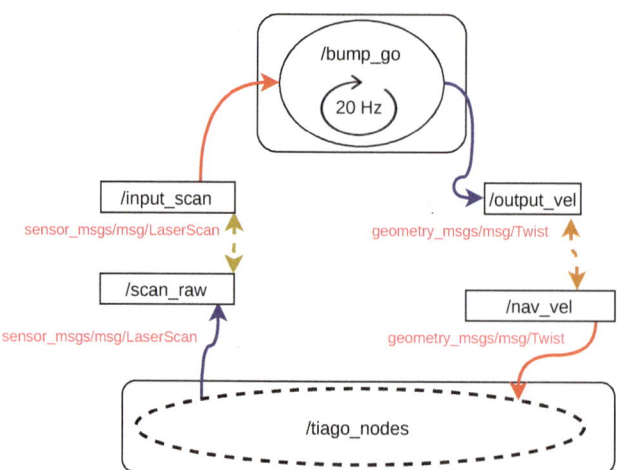

Figure 3.4: Computation Graph for *Bump and Go* project.

frequency we must publish it. You have to deal with this. *Engineers do not complain about problem conditions—they fix them.*

If we want our software to run on different robots, we must not specify specific topics for a robot. In our case, the topic that it subscribes to is /input_scan, and it publishes in /output_vel. These topics do not exist or correspond to those of our simulated robot. When executing it (at deployment), we will remap the ports to connect them to the real topics of the specific robot.

Let's expand upon this point here. Why are we using remaps instead of passing the name of the topics as parameters? Well, it is an alternative that many ROS 2 developers advocate. Perhaps this alternative is more convenient when a node does not always have the same subscribers/publishers, and this can only be specified in a YAML file of configuration parameters.

A good approach is that if the number of publishers and subscribers in a node is known, use generic topic names, like the ones used in this example, and perform a remap. It may even be better to use common topic names (/cmd_vel is a common topic for many robots). A seasoned ROS 2 programmer will read in the documentation what topics it uses, find out with a ros2 node info, and quickly make it work with remaps instead of looking for the correct parameters to be set up in the configuration files.

Although this book primarily uses C++, in this chapter we will provide two similar implementations, one in C++ and other in Python, each in different packages: br2_fsm_bumpgo_cpp and br2_fsm_bumpgo_py. Both are already in the workspace created in previous chapters and the annex to this book. Let's start with the C++ implementation.

3.3 *BUMP AND GO* IN C++

The br2_fsm_bumpgo_cpp package has the following structure:

```
Package br2_fsm_bumpgo_cpp

    br2_fsm_bumpgo_cpp
    ├── CMakeLists.txt
    ├── include
    │   └── br2_fsm_bumpgo_cpp
    │       └── BumpGoNode.hpp
    ├── launch
    │   └── bump_and_go.launch.py
    ├── package.xml
    └── src
        ├── br2_fsm_bumpgo_cpp
        │   └── BumpGoNode.cpp
        └── bumpgo_main.cpp
```

The usual way for nodes to be implemented as classes that inherit from `rclcpp::Node`, separating declaration and definition, within a namespace that matches the package name. In our case, the definition (BumpGoNode.cpp) will be in src/br2_fsm_bumpgo_cpp, and the header (BumpGoNode.hpp) will be in include/br2_fsm_bumpgo_cpp. In this way, we separate the implementation of the programs from the implementation of the nodes. This strategy allows having several programs with different strategies for creating nodes. The main program, whose function is to instantiate the node and call to the **spin()** function, is in src/bumpgo_main.cpp. We have also included a launcher (launch/bump_and_go.launch.py) to facilitate its execution.

In this book, we will analyze partial pieces of the code of a package, focusing on different concrete aspects to teach interesting concepts. We will not exhaustively show all the code since the reader has it available in his workspace, the repository, and the annexes.

3.3.1 Execution Control

The node execution model consists of calling the `control_cycle` method at a frequency of 20 Hz. For this, we declare a timer and start it in the constructor to call the `control_cycle` method every 50 ms. The control logic, implemented with an FSM, will publish the commands in speeds.

```cpp
include/bump_go_cpp/BumpGoNode.hpp

class BumpGoNode : public rclcpp::Node
{
  ...
private:
  void scan_callback(sensor_msgs::msg::LaserScan::UniquePtr msg);
  void control_cycle();

  rclcpp::Publisher<geometry_msgs::msg::Twist>::SharedPtr vel_pub_;
  rclcpp::Subscription<sensor_msgs::msg::LaserScan>::SharedPtr scan_sub_;
  rclcpp::TimerBase::SharedPtr timer_;

  sensor_msgs::msg::LaserScan::UniquePtr last_scan_;
};
```

Look at the detail of the laser callback header. We have used `UniquePtr` (an alias for `std::unique_ptr`) instead of `SharedPtr`, as we have seen so far. The Callbacks in ROS 2 can have different signatures, depending on the needs. These are different alternatives for the callbacks:

```
1. void scan_callback(const sensor_msgs::msg::LaserScan & msg);
2. void scan_callback(sensor_msgs::msg::LaserScan::UniquePtr msg);
3. void scan_callback(sensor_msgs::msg::LaserScan::SharedConstPtr msg);
4. void scan_callback(const sensor_msgs::msg::LaserScan::SharedConstPtr & msg);
5. void scan_callback(sensor_msgs::msg::LaserScan::SharedPtr msg);
```

Some other signatures allow us to obtain information about the message (timestamp in origin and destination, and identifier of the sender) and even the serialized message, but that is only used in very specialized cases.

Up to this point, we had used signature 1, but now we use signature 2. Check out the implementation of the laser callback at `scan_callback`. Instead of making a copy of the message (which could be computationally expensive for large messages) or sharing the pointer, we will acquire this message in property, and we will store the reference to the data in `last_scan_`. This way, rclcpp queues will no longer need to manage their lifecycle, saving time. We recommend using `UniquePtr` when possible to improve the performance of your nodes.

```
src/bump_go_cpp/BumpGoNode.cpp

BumpGoNode::BumpGoNode()
: Node("bump_go")
{
  scan_sub_ = create_subscription<sensor_msgs::msg::LaserScan>(
    "input_scan", rclcpp::SensorDataQoS(),
    std::bind(&BumpGoNode::scan_callback, this, _1));

  vel_pub_ = create_publisher<geometry_msgs::msg::Twist>("output_vel", 10);
  timer_ = create_wall_timer(50ms, std::bind(&BumpGoNode::control_cycle, this));
}

void
BumpGoNode::scan_callback(sensor_msgs::msg::LaserScan::UniquePtr msg)
{
  last_scan_ = std::move(msg);
}

void
BumpGoNode::control_cycle()
{
  // Do nothing until the first sensor read
  if (last_scan_ == nullptr)
    return;

  vel_pub_->publish(...);
}
```

Another noteworthy detail about this constructor is that the publication use the default QoS, which is reliable + volatile. In case of subscriptions, we will use `rclcpp::SensorDataQoS()` (a packed QoS definition using best effort, volatile, and appropiate queue size for sensors).

As a general rule, for a communication to be compatible, the quality of service of the publisher should be reliable, and it is the subscriber who can choose to relax it to be the best effort. When creating sensor drivers, publishing their readings using `rclcpp::SensorDataQoS()` is not a good idea because if a subscriber requires reliable QoS and publisher is best effort, communication will fail.

Finally, the first thing to do in `control_cycle` is to check if `last_scan_` is valid. This method may be executed before the first message arrives with a laser scan. In this case, this iteration is skipped.

3.3.2 Implementing an FSM

Implementing an FSM in a C++ class is not complicated. It is enough to have a member variable `state_` that stores the current state, which we can encode as a constant or an enum. In addition, it is helpful to have a variable `state_ts_` that indicates when transit to the current state, allowing to transit from states using timeouts.

```
include/bump_go_cpp/BumpGoNode.hpp

class BumpGoNode : public rclcpp::Node
{
  ...
private:
  void control_cycle();

  static const int FORWARD = 0;
  static const int BACK = 1;
  static const int TURN = 2;
  static const int STOP = 3;
  int state_;
  rclcpp::Time state_ts_;
};
```

Remember that the control logic is in method `control_cycle`, which runs at 20 Hz. This method must not contain any infinite loops or porlonged pauses.

Control logic is typically implemented with a `switch` statement, with a state in each case. In the following code, we have only shown the case of the `FORWARD` state. There is also a structure in this case: first, the output computation in the current state (setting speeds to publish) and then check every transition condition. If any returns true (the condition is met), the `state_` is set to the new state and `state_ts_` is updated.

When declaring a message type variable, all its fields are set by default to their default value, or 0 or empty depending on their type. That is why in the complete code, we only assign the field that is not 0.

```
src/bump_go_cpp/BumpGoNode.cpp
```

```cpp
BumpGoNode::BumpGoNode()
: Node("bump_go"),
  state_(FORWARD)
{
  ...

  state_ts_ = now();
}

void
BumpGoNode::control_cycle()
{
  switch (state_) {
    case FORWARD:

      // Do whatever you should do in this state.
      // In this case, set the output speed.

      // Checking the condition to go to another state in the next iteration
      if (check_forward_2_stop())
        go_state(STOP);
      if (check_forward_2_back())
        go_state(BACK);

      break;
      ...
  }
}

void
BumpGoNode::go_state(int new_state)
{
  state_ = new_state;
  state_ts_ = now();
}
```

Look at three methods with interesting code from the implementation point of view. The first is the code of the forward → back transition that checks if there is an obstacle in front of the robot. As we said before, this is done by accessing the central element of the vector that contains the distances in the laser reading:

```
src/bump_go_cpp/BumpGoNode.cpp
```

```cpp
bool
BumpGoNode::check_forward_2_back()
{
  // going forward when detecting an obstacle
  // at 0.5 meters with the front laser read
  size_t pos = last_scan_->ranges.size() / 2;
  return last_scan_->ranges[pos] < OBSTACLE_DISTANCE;
}
```

The second interesting snippet is the transition from forward → stop when the last laser read is considered too old. The now method of rclcpp::Node returns the current time as an rclcpp::Time. From the time that is in the header of the last reading, we can create another rclcpp::Time. Its difference is a rclcpp::Duration. To make comparisons, we can use its seconds method, which returns the time in seconds as a double, or we can, as we have done, directly compare it with another rclcpp::Duration.

```
src/bump_go_cpp/BumpGoNode.cpp

bool
BumpGoNode::check_forward_2_stop()
{
  // Stop if no sensor readings for 1 second
  auto elapsed = now() - rclcpp::Time(last_scan_->header.stamp);
  return elapsed > SCAN_TIMEOUT;  // SCAN_TIMEOUT is set to 1.0
}
```

The last snippet is similar to the previous one, but now we take advantage of having the `state_ts_` variable updated, and we can transition from `back` → `turn` after 2 s.

```
src/bump_go_cpp/BumpGoNode.cpp

bool
BumpGoNode::check_back_2_turn()
{
  // Going back for 2 seconds
  return (now() - state_ts_) > BACKING_TIME;
}
```

3.3.3 Running the Code

So far, we have limited ourselves to the class that implements the `BumpGoNode` node. Now, we have to see where we create an object of this class to execute it. We do this in the main program that creates a node and passes it to a blocking call to `rclcpp::spin` that will manage the messages and timer events calling to their callbacks.

```
src/bumpgo_main.cpp

int main(int argc, char * argv[])
{
  rclcpp::init(argc, argv);

  auto bumpgo_node = std::make_shared<br2_fsm_bumpgo_cpp::BumpGoNode>();
  rclcpp::spin(bumpgo_node);

  rclcpp::shutdown();

  return 0;
}
```

Now, run the program. Open a terminal to run the simulator:

```
$ ros2 launch br2_tiago sim.launch.py
```

Put some obstacles, a box, for example. Next, open another terminal and run the program, taking into account that there are arguments to specify in the command line:

- Remap `input_scan` to `/scan_raw`, and `ouput_vel` to `/nav_vel` (`-r` option).

- When using a simulator, set the `use_sim_time` parameter to `true`. This causes the time to be taken from the topic `/clock`, published by the simulator, instead of using the system clock.

```
$ ros2 run br2_fsm_bumpgo_cpp bumpgo --ros-args -r output_vel:=/nav_vel -r
input_scan:=/scan_raw -p use_sim_time:=true
```

See how the robot moves forward until it detects an obstacle then does an avoidance maneuver.

Because it is tedious to put so many remapping arguments in the command line, we have created a launcher that specifies the necessary arguments and remaps to the node.

```
launch/bump_and_go.launch.py

bumpgo_cmd = Node(package='br2_fsm_bumpgo_cpp',
    executable='bumpgo',
    output='screen',
    parameters=[{
        'use_sim_time': True
    }],
    remappings=[
        ('input_scan', '/scan_raw'),
        ('output_vel', '/nav_vel')
    ])
```

Use this launcher instead of the last `ros2 run`, only by typing:

```
$ ros2 launch br2_fsm_bumpgo_cpp bump_and_go.launch.py
```

3.4 *BUMP AND GO* BEHAVIOR IN PYTHON

In addition to C++, Python is one of the languages officially supported in ROS 2 through the rclpy client library. This section will reproduce what we have done in the previous section, but with Python. Verify by comparison the differences and similarities in the development of both languages. Also, once the principles of ROS 2 have been explained throughout the previous chapters, the reader will recognize the elements of ROS 2 in Python code, as the principles are the same.

Although we provided the complete package, if we had wanted to create a package from scratch, we could have used the `ros2 pkg` command to create a skeleton.

```
$ ros2 pkg create --build-type ament_python br2_fsm_bumpgo_py --dependencies
sensor_msgs geometry_msgs
```

As it is a ROS 2 package, there is still a `package.xml` similar to the C++ version, but there is no longer a `CMakeLists.txt`, but a `setup.cfg` and `setup.py`, typical of Python packages that use distutils[2].

At the root of this package, there is a homonymous directory that only has a file `__init__.py` which indicates that there will be files with Python code. Let's create the file `bump_go_main.py` there. While in C++, it is common and convenient to separate the source code into several files. In this case, everything is in the same file.

---

[2]`https://docs.python.org/3/library/distutils.html`

3.4.1 Execution Control

As in the previous example, we will first show the code ignoring the details of the behavior, only those related to the ROS 2 concepts to handle:

```python
bump_go_py/bump_go_main.py

import rclpy

from rclpy.duration import Duration
from rclpy.node import Node
from rclpy.qos import qos_profile_sensor_data
from rclpy.time import Time

from geometry_msgs.msg import Twist
from sensor_msgs.msg import LaserScan


class BumpGoNode(Node):
    def __init__(self):
        super().__init__('bump_go')

        ...

        self.last_scan = None
        self.scan_sub = self.create_subscription(
            LaserScan,
            'input_scan',
            self.scan_callback,
            qos_profile_sensor_data)

        self.vel_pub = self.create_publisher(Twist, 'output_vel', 10)
        self.timer = self.create_timer(0.05, self.control_cycle)

    def scan_callback(self, msg):
        self.last_scan = msg

    def control_cycle(self):
        if self.last_scan is None:
            return

        out_vel = Twist()

        # FSM

        self.vel_pub.publish(out_vel)

def main(args=None):
    rclpy.init(args=args)

    bump_go_node = BumpGoNode()

    rclpy.spin(bump_go_node)

    bump_go_node.destroy_node()
    rclpy.shutdown()

if __name__ == '__main__':
    main()
```

Recall that the goal is to create a node that subscribes to the laser readings and issues speed commands. The control cycle executes at 20 Hz to calculate the robot control based on the last reading received. Therefore, our code will have a subscriber, a publisher, and a timer.

This code is similar to the one developed in C++: define a class that inherits from Node, and in the main, it is instantiated and called spin with it. Let's see some details:

- Inheriting from Node, we call the base class constructor to assign the node name. The Node class and all associated data types (Time, Duration, QoS,...) are in rclpy, imported at startup, and these items separately.

- The types of messages are also imported, as seen in the initial part.

- We create the publisher, the subscriber, and the timer in the constructor. Note that the API is similar to C++. Also, in Python, we can access predefined qualities of service (`qos_profile_sensor_data`).

- In the callback of the laser messages, we store the last message received in the variable `self.last_scan`, which was initialized to None in the constructor. In this way, verify in the control cycle (`control_cycle`) that no laser reading has reached us.

3.4.2   Implementing the FSM

The direct translation of the FSM in C++ from the previous section to Python has nothing interesting. The only difference is that to obtain the current time, we have to ask for the clock first through the `get_clock` method:

```
bump_go_py/bump_go_main.py

class BumpGoNode(Node):
    def __init__(self):
        super().__init__('bump_go')

        self.FORWARD = 0
        self.BACK = 1
        self.TURN = 2
        self.STOP = 3
        self.state = self.FORWARD
        self.state_ts = self.get_clock().now()

    def control_cycle(self):

        if self.state == self.FORWARD:
          out_vel.linear.x = self.SPEED_LINEAR

            if self.check_forward_2_stop():
              self.go_state(self.STOP)
            if self.check_forward_2_back():
              self.go_state(self.BACK)

        self.vel_pub.publish(out_vel)

    def go_state(self, new_state):
        self.state = new_state
        self.state_ts = self.get_clock().now()
```

Perhaps the most remarkable aspect in this code, similar to its version in C++, is the treatment of time and durations:

```
bump_go_py/bump_go_main.py
    def check_forward_2_back(self):
        pos = round(len(self.last_scan.ranges) / 2)
        return self.last_scan.ranges[pos] < self.OBSTACLE_DISTANCE

    def check_forward_2_stop(self):
        elapsed = self.get_clock().now() - Time.from_msg(self.last_scan.header.stamp)
        return elapsed > Duration(seconds=self.SCAN_TIMEOUT)

    def check_back_2_turn(self):
        elapsed = self.get_clock().now() - self.state_ts
        return elapsed > Duration(seconds=self.BACKING_TIME)
```

- The `Time.from_msg` function allows to create a `Time` object from the timestamp of a message.

- The current time is obtained with Node's `get_clock().now()` method.

- The operation between time has as a result an object of type `Duration`, which can be compared with another object of type `Duration`, such as `Duration(seconds = self.BACKING_TIME)` that represents the duration of 2 seconds.

3.4.3 Running the Code

Let's see how to build and install the code in the workspace. First, Modify `setup.py` for our new program:

```
setup.py
import os
from glob import glob

from setuptools import setup

package_name = 'br2_fsm_bumpgo_py'

setup(
    name=package_name,
    version='0.0.0',
    packages=[package_name],
    data_files=[
        ('share/ament_index/resource_index/packages',
            ['resource/' + package_name]),
        ('share/' + package_name, ['package.xml']),
        (os.path.join('share', package_name, 'launch'), glob('launch/*.launch.py'))
    ],
    install_requires=['setuptools'],
    zip_safe=True,
    maintainer='fmrico',
    maintainer_email='fmrico@gmail.com',
    description='BumpGo in Python package',
    license='Apache 2.0',
    tests_require=['pytest'],
    entry_points={
        'console_scripts': [
            'bump_go_main = br2_fsm_bumpgo_py.bump_go_main:main'
        ],
    },
)
```

The important part right now is the `entry_points` argument. As shown in the code above, add the new program shown previously. With this, we can already build our package.

```
$ colcon build --symlink-install
```

In order to run the program, first launch the simulator by typing in the terminal:

```
$ ros2 launch br2_tiago sim.launch.py
```

Open another terminal, and run the program:

```
$ ros2 run br2_fsm_bumpgo_py bump_go_main --ros-args -r output_vel:=/nav_vel -r
input_scan:=/scan_raw -p use_sim_time:=true
```

We can also use a launcher similar to the one in the C++ version by just typing:

```
$ ros2 launch br2_fsm_bumpgo_py bump_and_go.launch.py
```

PROPOSED EXERCISES:

1. Modify the *Bump and Go* project so that the robot perceives an obstacle in front, on its left and right diagonal. Instead of always turning to the same side, it turns to the side with no obstacle.

2. Modify the *Bump and Go* project so that the robot turns exactly to the angle with no obstacles or the more far perceived obstacle. Try two approaches:

   • Open-loop: Calculate before turning time and speed to turn.
   • Closed-loop: Turns until a clear space in front is detected.

## CHAPTER 4

The TF Subsystem

O NE of the greatest hidden treasures in ROS is its geometric transformation subsystem TF (or TFs in short). This subsystem allows defining different reference axes (also called frames) and the geometric relationship between them, even when this relationship is constantly changing. Any coordinate in a frame can be recalculated to another frame without the need for tedious manual calculations.

In my experience teaching ROS courses, students who had to deal with similar calculations without TFs were always happy upon learning them.

Its importance in ROS is due to the need to model the parts and components of a robot geometrically. It has many applications in navigation and location, as well as manipulation. They have been used to position several cameras in a building or motion capture systems[1].

A robot perceives the environment through sensors placed somewhere on the robot and performs actions for which it needs to specify some spatial position. For instance:

- A distance sensor (laser or RGBD) generates a set of points (x, y, z) that indicate the detected obstacles.

- A robot moves its end effector by specifying a target position $(x, y, z, roll, pitch, yaw)$.

- A robot moves to a point (x, y, yaw) on a map.

All these coordinates are references to a frame. In a robot, there are multiple frames (for sensors, actuators, etc.). The relationship between these frames must be known to reason. For example, the coordinate of an obstacle detected by the laser on the arm reference axis to avoid it. Frames relationships are the displacement and rotation of a frame to another frame. Algebraically, this is done using homogeneous coordinates for the coordinates and RT transformation matrices for relations. Having the coordinates of a point P in frame A, this is P_A, we can calculate P_B in frame B using the transformation matrix $RT_{A \to B}$ as follows:

---

[1] https://github.com/MOCAP4ROS2-Project

$$P_B = RT_{A \to B} * P_A \tag{4.1}$$

$$
\begin{pmatrix} x_B \\ y_B \\ z_B \\ 1 \end{pmatrix} = \begin{pmatrix} R^{xx}_{A \to B} & R^{xy}_{A \to B} & R^{xz}_{A \to B} & T^{x}_{A \to B} \\ R^{yx}_{A \to B} & R^{yy}_{A \to B} & R^{yz}_{A \to B} & T^{y}_{A \to B} \\ R^{zx}_{A \to B} & R^{zy}_{A \to B} & R^{zz}_{A \to B} & T^{z}_{A \to B} \\ 0 & 0 & 0 & 1 \end{pmatrix} * \begin{pmatrix} x_A \\ y_A \\ z_A \\ 1 \end{pmatrix} \tag{4.2}
$$

In addition to the complexity of these operations, it is remarkable that these relationships are highly dynamic in an articulated robot. It would be an error to transform the points perceived by a sensor at time t using the transformation at $t + 0.01$ s if it varies dynamically at high speed.

ROS 2 implements the TF transform system (now called TF2, the second version) using two topics that receives transformations, as messages of type tf2_msgs/msg/TFMessage:

```
$ ros2 interface show tf2_msgs/msg/TFMessage

geometry_msgs/TransformStamped[] transforms
    std_msgs/Header header
    string child_frame_id
    Transform transform
        Vector3 translation
            float64 x
            float64 y
            float64 z
        Quaternion rotation
            float64 x 0
            float64 y 0
            float64 z 0
            float64 w 1
```

- **/tf** for transforms that vary dynamically, like the joints of a robot are specified here. By default, they are valid for a short time (10 s). For example, frames relation linked by motorized joints are published here.

- **/tf_static** for transforms that do not vary over time. This topic has a QoS transient_local, so any node that subscribes to this topic receives all the transforms published so far. Typically, the transforms published to this topic do not change over time, like the robot geometry.

The frames of a robot are organized as a tree of TFs, in which each TF *should* have at most one parent and can have several children. If this is not true, or several trees are not connected, the robot is not well modeled. By convention, there are several important axes:

- **base_footprint** is usually the root of a robot's TFs, and corresponds to the center of the robot on the ground. It is helpful to transform the information from the robot's sensors to this axis to relate them to each other.

- **base_link** is usually the child of /base_footprint, and is typically the center of the robot, already above ground level.

- **odom** is the parent frame of /base_footprint, and the transformation that relates them indicates the robot's displacement since the robot's motion started.

Figure 4.1 shows partially the TF tree of the simulated Tiago. If you would like to see the whole TF tree, launch the simulation and then type[2]:

```
$ ros2 run rqt_tf_tree rqt_tf_tree
```

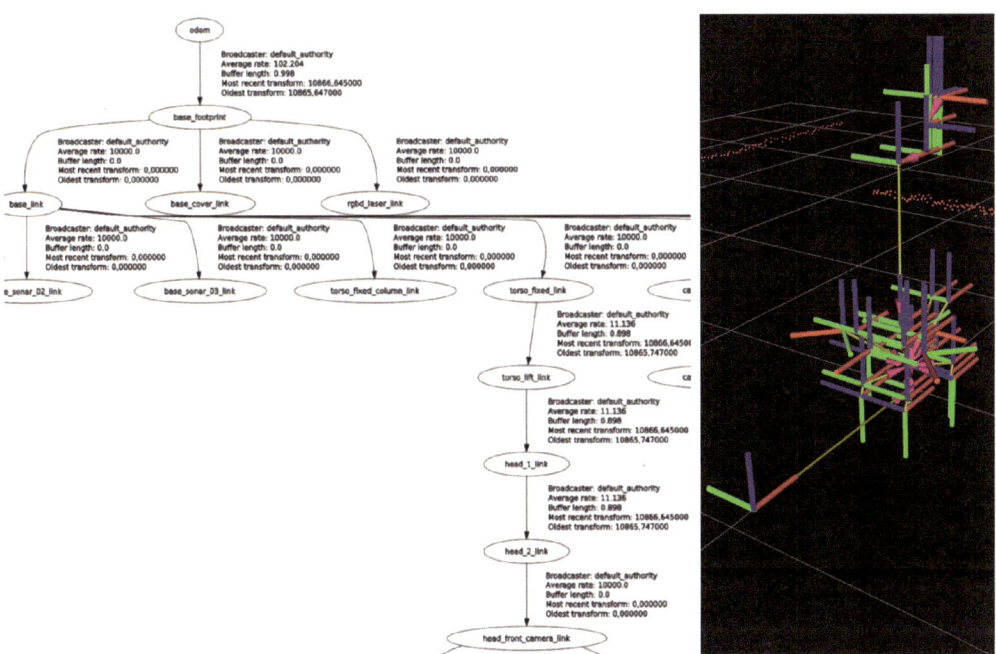

Figure 4.1: Portion of the TF tree of the simulated Tiago and the TF display in RViz2.

When a node wants to use this system, it does not subscribe directly to these topics but uses **TFListeners**, which are objects that update a buffer that stores all the latest published TFs, and that has an API that lets, for example:

- To know if there is a TF from one frame to another at time t.

- To know what is the rotation from frame A to frame B at time t.

- To ask to transform a coordinate that is in frame A and to frame B in an arbitrary time t.

The buffer may not contain just the TF at time t, but if it has an earlier and a later one, it performs the interpolation. Likewise, frames A and B may not be

---

[2]It is needed to have installed the package **ros-jazzy-rqt-tf-tree**

directly connected, but more frames are in between, performing the necessary matrix operations automatically.

Without going into much detail, for now, publishing a transform to a ROS 2 node is very straightforward. Just have a transform broadcaster and send transforms to the TF system:

```
geometry_msgs::msg::TransformStamped detection_tf;

detection_tf.header.frame_id = "base_footprint";
detection_tf.header.stamp = now();
detection_tf.child_frame_id = "detected_obstacle";
detection_tf.transform.translation.x = 1.0;

tf_broadcaster_->sendTransform(detection_tf);
```

Getting a transform is easy too. By having a TF buffer that a transform listener updates, we can ask for the geometric transformation from one frame to another. Not even these frames need to be directly connected. Any calculation is done transparently for the developer:

```
tf2_ros::Buffer tfBuffer;
tf2_ros::TransformListener tfListener(tfBuffer);

...

geometry_msgs::msg::TransformStamped odom2obstacle;
odom2obstacle = tfBuffer_.lookupTransform("odom", "detected_obstacle", tf2::TimePointZero);
```

The above code calculates odom → base_footprint → detected_obstacle automatically. The third argument of `lookupTransform` indicates the instant of time from which we want to obtain the transform. `tf2::TimePointZero` indicates the latest available. If we are transforming points of a laser, for example, we should use the timestamp that appears in the header of the laser message, because if a robot or the laser has moved since then, the transformation in another instant will not be exact (much can change in few milliseconds in a robot). Finally, be careful about asking for the transforms with `now()`, because it will not have information yet at this moment in time, and it cannot be extrapolated into the future, and an exception can be raised.

We can operate with transforms, multiplying them or calculating their inverse. From here, we will establish a nomenclature convention for our code. This will help us to operate with TFs:

- If an object represent a transformation from frame `origin` to frame `target`, we call it `origin2target`.

- If needed to multiply two TFs, as shown in Figure 4.2.

  1. We only can operate it if the frame names near operator * are equal. In this case, the frame names are equals (robot).

  2. The result frame id must be the outer part of the operators (odom from first operator and object from second).

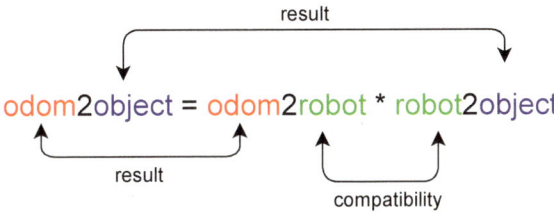

Figure 4.2: The mnemonic rule for naming and operating TFs. Based on their name, we can know if two TFs can be multiplied and the name of the resulting TF.

3. If we invert a TF (they are invertibles), we invert the frame ids in this name.

4.1 AN OBSTACLE DETECTOR THAT USES TF2

This section will analyze a project to see in practice the application of the concepts on TFs introduced in the previous sections.

This project makes the robot detect obstacles right in front of it using the laser sensor, as shown in Figure 4.3.

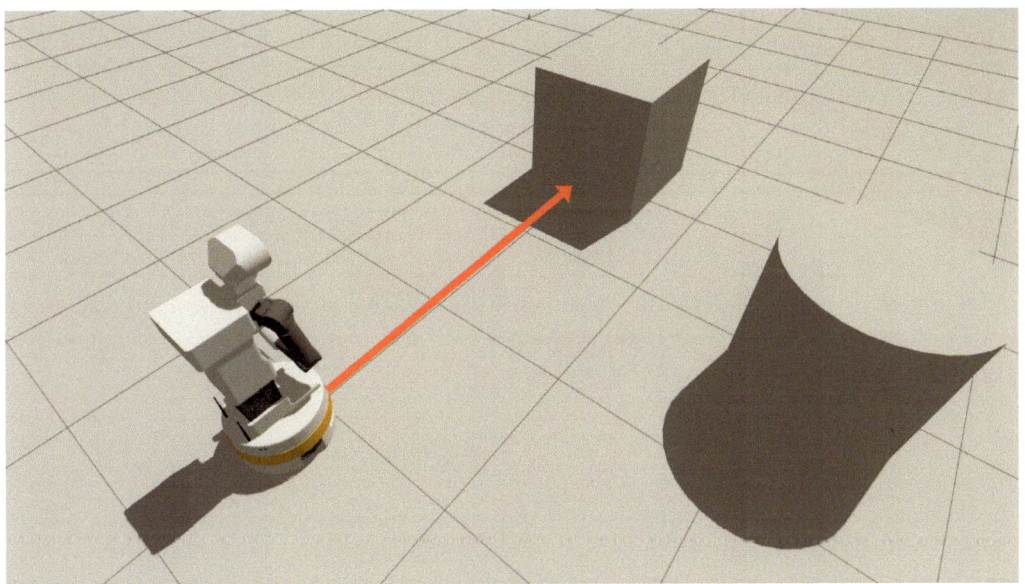

Figure 4.3: Robot Tiago detecting obstacles with the laser sensor. The red arrow highlights the obstacle detected with the center reading.

We will apply TFs concepts following a common practice in many ROS 2 packages to publish the perceptions as TFs. The advantage of doing this is that we can easily reason its position geometrically for any frame, even if it is not currently perceived.

We will not introduce a new **perception model**, but we will use the same one from the previous chapter: we will detect obstacles in front of the robot using the

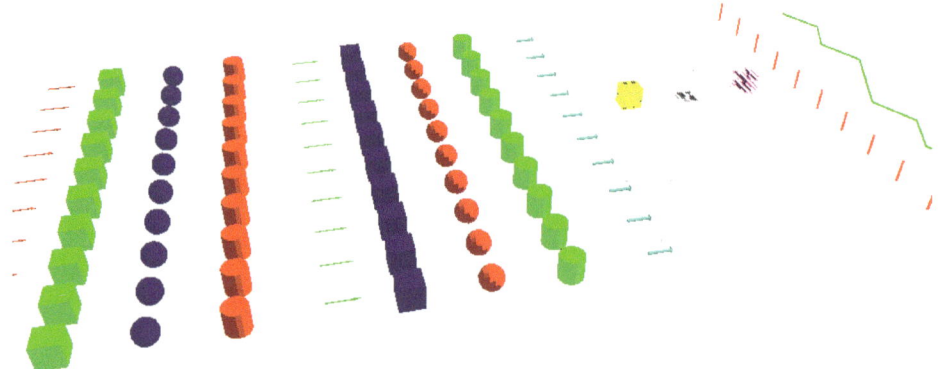

Figure 4.4: Visual markers available for visual debugging.

laser. We will use the same speed-based actuation model, although we will teleoperate the robot manually in this case.

In this project, apart from using the concepts about TFs, we will show a powerful debugging tool called *Visual Markers*[3], which allows us to publish 3D visual elements that can be viewed in RViz2 from a node. This mechanism allows us to show at a glance part of the internal state of the robot without limiting ourselves to the debugging messages that are generated with the macros `RCLCPP_*`. Markers include arrows, lines, cylinders, spheres, lines, shapes, text, and others in any size or color. Figure 4.4 shows an example of available markers.

4.2 COMPUTATION GRAPH

The Computation Graph of our application is shown in the Figure 4.5.

The node uses a laser sensor of the simulated robot at the `scan_raw` topic. The detection node subscribes to the laser topic and publishes the transform in the ROS 2 TF subsystem. Our node subscribes to `/input_scan`, so we will have to remap from `/scan_raw`.

We will create a node `/obstacle_monitor` that reads the transform corresponding to the detection and shows in console its position with respect to the general frame of the robot, `base_footprint`.

The node `/obstacle_monitor` publishes also a visual marker. In our case, we will publish a red arrow that connects the robot's base with the frame's position of the obstacle that we are publishing.

In this project, we will make two versions: a basic one and an improved one. The reason is to see a small detail about the use of TFs that significantly impacts the final result, as we will explain later.

---

[3]`http://wiki.ros.org/rviz/DisplayTypes/Marker`

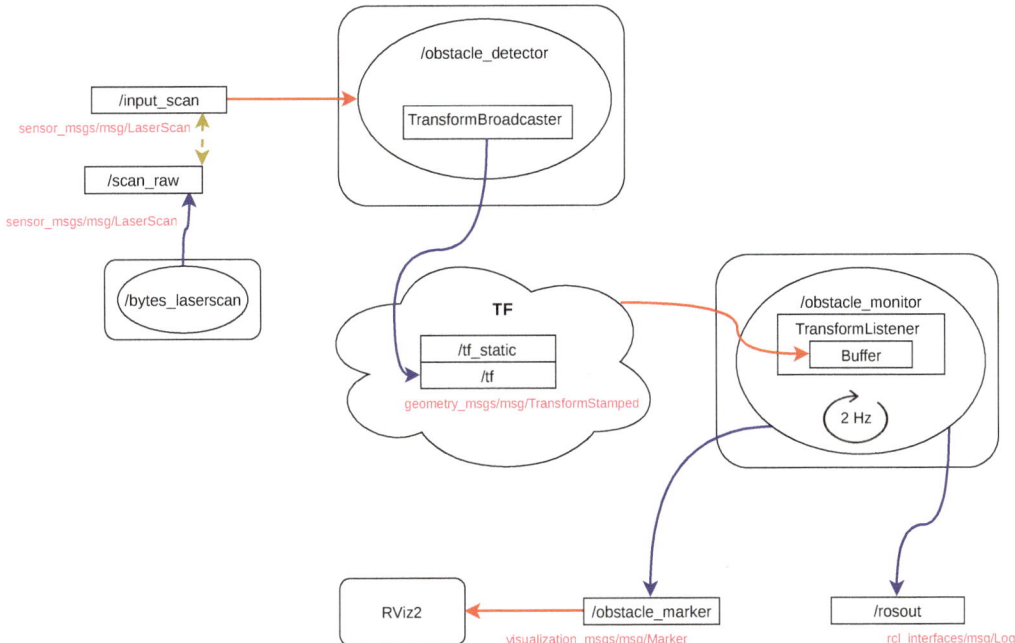

Figure 4.5: Computation Graph of the exercise. The **/obstacle_detector** node collaborates with the **/obstacle_monitor** node using the TF subsystem.

4.3 BASIC DETECTOR

We use the same package for both versions. The structure of the package can be seen in the following box:

```
Package br2_tf2_detector

br2_tf2_detector
├── CMakeLists.txt
├── include
│   └── br2_tf2_detector
│       ├── ObstacleDetectorImprovedNode.hpp
│       ├── ObstacleDetectorNode.hpp
│       └── ObstacleMonitorNode.hpp
├── launch
│   ├── detector_basic.launch.py
│   └── detector_improved.launch.py
├── package.xml
└── src
    ├── br2_tf2_detector
    │   ├── ObstacleDetectorImprovedNode.cpp
    │   ├── ObstacleDetectorNode.cpp
    │   └── ObstacleMonitorNode.cpp
    ├── detector_improved_main.cpp
    └── detector_main.cpp
```

We will ignore in this section the files that contain the word "Improved" in the name. We will see them in the next section.

The reader can see how the package structure is similar to the previous chapter. The nodes are separated in their declaration and definition, in directories whose name matches the package. In addition, everything will be defined within a namespace that matches the package's name. This package will take a small step forward in this structure: now, we will compile the nodes as a dynamic library linked by the executables. Perhaps in this project we will not notice the difference, but we save space, it is more convenient, and it could allow (it is not the case) to export it to other packages. The name of the library will be the name of the package (${PROJECT_NAME}), as usual when creating a support library in a package. Let's see what this looks like in the CMakeLists.txt file:

```
include/br2_tf2_detector/ObstacleDetectorNode.hpp

project(br2_tf2_detector)

find_package(...)
...

set(dependencies
...
)

include_directories(include)

add_library(${PROJECT_NAME} SHARED
  src/br2_tf2_detector/ObstacleDetectorNode.cpp
  src/br2_tf2_detector/ObstacleMonitorNode.cpp
  src/br2_tf2_detector/ObstacleDetectorImprovedNode.cpp
)
ament_target_dependencies(${PROJECT_NAME} ${dependencies})

add_executable(detector src/detector_main.cpp)
ament_target_dependencies(detector ${dependencies})
target_link_libraries(detector ${PROJECT_NAME})

add_executable(detector_improved src/detector_improved_main.cpp)
ament_target_dependencies(detector_improved ${dependencies})
target_link_libraries(detector_improved ${PROJECT_NAME})

install(TARGETS
  ${PROJECT_NAME}
  detector
  detector_improved
  ARCHIVE DESTINATION lib
  LIBRARY DESTINATION lib
  RUNTIME DESTINATION lib/${PROJECT_NAME}
)
```

Note that now it is needed to add a target_link_libraries statement and install the library in the same place as the executables. When specifying the files of each executable, it is no longer necessary to specify more than the main cpp program file.

4.3.1 Obstacle Detector Node

Analyze the obstacle detector node. Its execution follows an event-oriented model rather than an iterative one. Every message the node receives will produce an output, so it makes sense that the node's logic resides in the laser callback.

```
include/br2_tf2_detector/ObstacleDetectorNode.hpp
```

```cpp
class ObstacleDetectorNode : public rclcpp::Node
{
public:
  ObstacleDetectorNode();

private:
  void scan_callback(sensor_msgs::msg::LaserScan::UniquePtr msg);

  rclcpp::Subscription<sensor_msgs::msg::LaserScan>::SharedPtr scan_sub_;
  std::shared_ptr<tf2_ros::StaticTransformBroadcaster> tf_broadcaster_;
};
```

Since the node must publish transforms to the TF subsystem, we declare a StaticTransformBroadcaster, that publish in /tf_static. We could also declare a TransformBroadcaster that publish in /tf. Apart from the durability QoS, the difference is that we want transforms to persist beyond the 10 s by default of non-static transforms.

We use a shared_ptr for tf_broadcaster_, since its constructor requires an rclcpp::Node*, and we will not have it until we are already inside the constructor[4]:

```
src/br2_tf2_detector/ObstacleDetectorNode.hpp
```

```cpp
ObstacleDetectorNode::ObstacleDetectorNode()
: Node("obstacle_detector")
{
  scan_sub_ = create_subscription<sensor_msgs::msg::LaserScan>(
    "input_scan", rclcpp::SensorDataQoS(),
    std::bind(&ObstacleDetectorNode::scan_callback, this, _1));

  tf_broadcaster_ = std::make_shared<tf2_ros::TransformBroadcaster>(*this);
}
```

The tf_broadcaster_ object manages the publication of static TFs. The message type of a TF is geometry_msgs/msg/TransformStamped. Let's see how it is used:

```
src/br2_tf2_detector/ObstacleDetectorNode.hpp
```

```cpp
void
ObstacleDetectorNode::scan_callback(sensor_msgs::msg::LaserScan::UniquePtr msg)
{
  double dist = msg->ranges[msg->ranges.size() / 2];

  if (!std::isinf(dist)) {
    geometry_msgs::msg::TransformStamped detection_tf;

    dotoction_tf.header = msg->header;
    detection_tf.child_frame_id = "detected_obstacle";
    detection_tf.transform.translation.x = msg->ranges[msg->ranges.size() / 2];

    tf_broadcaster_->sendTransform(detection_tf);
  }
}
```

- The header of the output message will be the header of the input laser message. We will do this because the timestamp must be when the sensory reading was taken. If we used now(), depending on the latency in the messages and the

---

[4]In fact, some C++ developers recommend avoiding using this in constructors, as the object has not completely initialized until the constructor finishes.

load of the computer, the transform would not be precise, and synchronization errors could occur.

The `frame_id` is the source frame (or parent frame) of the transformation, already in this header. In this case, it is the sensor frame since the perceived coordinates of the object are in this frame.

- The `child_frame_id` field is the id of the new frame that we are going to create, and that represents the perceived obstacle.

- The transform field contains a translation and a rotation applied in this order, from the parent frame to the child frame that we want to create. Since the X-axis of the laser frame is aligned with the laser beam that we are measuring, the translation in X is the distance read.

  Rotation refers to the rotation of the frame after translation is applied. As this value is not relevant here (detection is a point) we use the default quaternion values $(0, 0, 0, 1)$ set by the message constructor.

- Finally, use the `sendTransform()` method of `tf_broadcaster_` to send the transform to the TF subsystem.

4.3.2 Obstacle Monitor Node

The `/obstacle_monitor` node extracts the transform to the detected object from the TFs system and shows it to the user in two ways:

- The standard output on the console indicates where the obstacle is with respect to the robot at all times, even if it is no longer being detected.

- Using a visual marker, specifically an arrow, which starts from the robot toward the obstacle that was detected.

Analyze the header to see what elements this node has:

```
include/br2_tf2_detector/ObstacleMonitorNode.hpp

class ObstacleMonitorNode : public rclcpp::Node
{
public:
  ObstacleMonitorNode();

private:
  void control_cycle();
  rclcpp::TimerBase::SharedPtr timer_;

  tf2::BufferCore tf_buffer_;
  tf2_ros::TransformListener tf_listener_;

  rclcpp::Publisher<visualization_msgs::msg::Marker>::SharedPtr marker_pub_;
};
```

- The execution model of this node is iterative, so we declare `timer_` and its callback `control_cycle`.

- To access the TF system, use a `tf2_ros::TransformListener` that update the buffer `tf_buffer_` to which we can make the queries we need.

- We only need one publisher for visual markers.

In the case of the class definition, we ignore the part dedicated to visual markers, for now, showing only the part related to TFs.

```
src/br2_tf2_detector/ObstacleMonitorNode.cpp

 1   ObstacleMonitorNode::ObstacleMonitorNode()
 2   : Node("obstacle_monitor"),
 3     tf_buffer_(),
 4     tf_listener_(tf_buffer_)
 5   {
 6     marker_pub_ = create_publisher<visualization_msgs::msg::Marker>(
 7       "obstacle_marker", 1);
 8
 9     timer_ = create_wall_timer(
10       500ms, std::bind(&ObstacleMonitorNode::control_cycle, this));
11   }
12
13   void
14   ObstacleMonitorNode::control_cycle()
15   {
16     geometry_msgs::msg::TransformStamped robot2obstacle;
17
18     try {
19       robot2obstacle = tf_buffer_.lookupTransform(
20         "base_footprint", "detected_obstacle", tf2::TimePointZero);
21     } catch (tf2::TransformException & ex) {
22       RCLCPP_WARN(get_logger(), "Obstacle transform not found: %s", ex.what());
23       return;
24     }
25
26     double x = robot2obstacle.transform.translation.x;
27     double y = robot2obstacle.transform.translation.y;
28     double z = robot2obstacle.transform.translation.z;
29     double theta = atan2(y, x);
30
31     RCLCPP_INFO(get_logger(), "Obstacle detected at (%lf m, %lf m, , %lf m) = %lf rads",
32       x, y, z, theta);
33   }
```

- Notice how `tf_listener_` is initialized by simply specifying the buffer to update. Later, the queries will be made directly to the buffer.

- We observe that the control loop runs at 2 Hz, showing us information with `RCLCPP_INFO` (to `/ros_out` and stdout).

- The most relevant function is `lookupTransform`, which calculates the geometric transformation from one frame to another, even if there is no direct relationship. We can specify a specific timestamp or, on the contrary, we want the last one available by indicating `tf2::TimePointZero`. This call can throw an exception if it does not exist, or we require a transform on a timestamp in the future, so a `try/catch` should be used to handle possible errors.

- Note that the TF we published in `ObstacleDetectorNode` was `base_laser_link` → `detected_obstacle`, and now we are requiring `base_footprint` → `detected_obstacle`. As the robot is well modeled and the geometric relationship between `base_laser_link` and `base_footprint` can be calculated, there will be no problem for `lookupTransform` to return the correct information.

Let's see the part related to the generation of the visual marker. The goal is to show the coordinates of the obstacle to the robot on the screen and show a geometric shape in RViz2 that allows us to debug the application visually. In this case, it will be a red arrow from the robot to the obstacle. To do this, create an `visualization_msgs/msg/Marker` message and fill in its fields to obtain this arrow:

```
src/br2_tf2_detector/ObstacleMonitorNode.cpp

  visualization_msgs::msg::Marker obstacle_arrow;
  obstacle_arrow.header.frame_id = "base_footprint";
  obstacle_arrow.header.stamp = now();
  obstacle_arrow.type = visualization_msgs::msg::Marker::ARROW;
  obstacle_arrow.action = visualization_msgs::msg::Marker::ADD;
  obstacle_arrow.lifetime = rclcpp::Duration(1s);

  geometry_msgs::msg::Point start;
  start.x = 0.0;
  start.y = 0.0;
  start.z = 0.0;
  geometry_msgs::msg::Point end;
  end.x = x;
  end.y = y;
  end.z = z;
  obstacle_arrow.points = {start, end};

  obstacle_arrow.color.r = 1.0;
  obstacle_arrow.color.g = 0.0;
  obstacle_arrow.color.b = 0.0;
  obstacle_arrow.color.a = 1.0;

  obstacle_arrow.scale.x = 0.02;
  obstacle_arrow.scale.y = 0.1;
  obstacle_arrow.scale.z = 0.1;
```

In the reference document[5] for visual markers is documented the meaning of every field for every type of marker. In the case of an arrow, the points field will be filled with the starting point $(0, 0, 0)$ and the ending point corresponding to the detection, both in `base_footprint`. Do not forget to assign a color, especially the `alpha`, since we will not see anything if we let it be 0, the default value.

4.3.3 Running the Basic Detector

We instantiate both in the same process to test our nodes, and we use a `SingleThreadedExecutor`. That would be enough to spin both:

---

[5]`://wiki.ros.org/rviz/DisplayTypes/Marker`

```
src/br2_tf2_detector/ObstacleMonitorNode.cpp

int main(int argc, char * argv[]) {
  rclcpp::init(argc, argv);

  auto obstacle_detector = std::make_shared<br2_tf2_detector::ObstacleDetectorNode>();
  auto obstacle_monitor = std::make_shared<br2_tf2_detector::ObstacleMonitorNode>();

  rclcpp::executors::SingleThreadedExecutor executor;
  executor.add_node(obstacle_detector->get_node_base_interface());
  executor.add_node(obstacle_monitor->get_node_base_interface());

  executor.spin();

  rclcpp::shutdown();
  return 0;
}
```

Follow the next commands to test our nodes:

```
# Terminal 1: The Tiago simulation
$ ros2 launch br2_tiago sim.launch.py world:=empty
```

```
# Terminal 2: Launch our nodes
$ ros2 launch br2_tf2_detector detector_basic.launch.py
```

```
# Terminal 3: Keyboard teleoperation
$ ros2 run teleop_twist_keyboard teleop_twist_keyboard --ros-args -r
cmd_vel:=/key_vel
```

```
# Terminal 4: RViz2
$ ros2 run rviz2 rviz2 --ros-args -p use_sim_time:=true
```

In Gazebo, add an obstacle in front of the robot. Start watching in the terminal the information about the detection. In Rviz2, change the fixed frame to odom. Add a Markers display to RViz2 specifying the topic that we have created to publish the visual marker. Also, add the TF Display if it is not added yet. Figure 4.6 shows the TF to the obstacle and also the red arrow.

Do a quick exercise: with the teleoperator, move the robot forward and to the side, so it no longer perceives the obstacle. Keep moving the robot and realize that the information returned by lookupTransform is no longer correct. It continues to indicate that the obstacle is ahead, although this is no longer true. What has happened? We probably wanted the arrow to point to the obstacle position, but now the arrow travels fixed with the robot.

Let's explain it with a diagram in Figure 4.7. As long as the robot perceives the obstacle, the transform requested (pink arrow) is correct. It is a transform from the robot's laser to the obstacle. When we stop updating the transform (thick blue arrow) because the obstacle is gone, the transform continues to exist. If we move the robot, lookupTransform keeps returning the last valid transform: in front fo the robot. This makes the visual marker wrong as well. The following section presents a strategy to fix this undesirable situation.

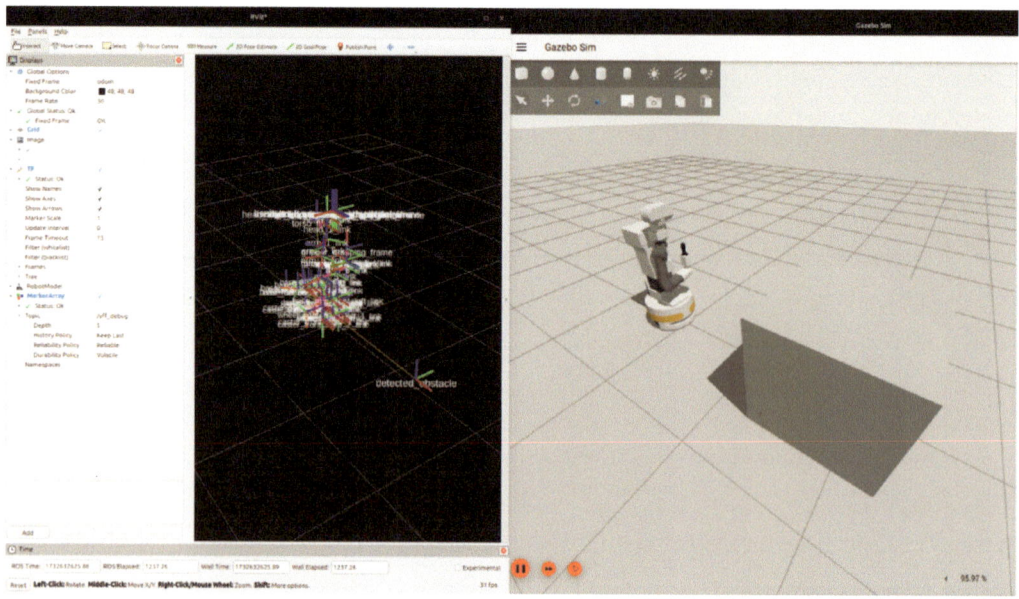

Figure 4.6: Visualization in RViz2 of the TF corresponding to the detection, and the red arrow marker published to visualize the detection.

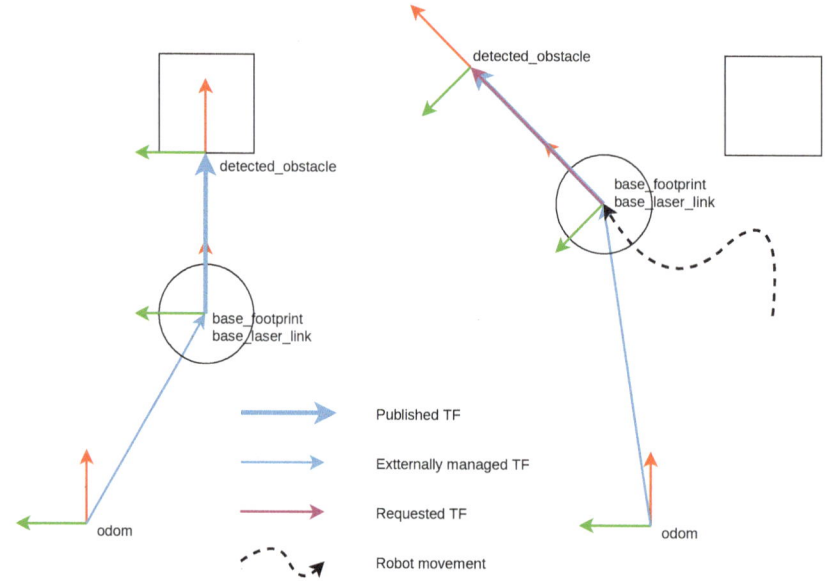

Figure 4.7: Diagram showing the problem when publishing TFs in the local frame. When the robot moves, the TF no longer represents the right obstacle position.

4.4 IMPROVED DETECTOR

The solution is to publish the detection TF in a fixed frame that is not affected by the robot's movement, for example, `odom` (or `map` if your robot is navigating). If we do it like this, when we require the transform `base_footprint` → `detected_obstacle` (pink arrow), this transform will be calculated taking into account the movement of the robot, collected in the transformation `odom` → `base_footprint`. This calculation is visualized in the diagram in Figure 4.8.

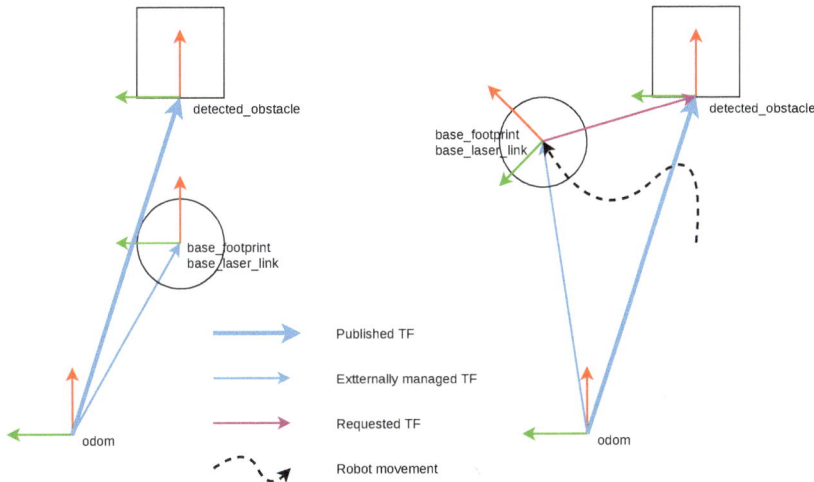

Figure 4.8: Diagram showing how to correctly maintain the obstacle position, by publishing the TF in a fixed frame. The calculated TF (thick blue arrow) takes into account the robot displacement.

`ObstacleDetectorImprovedNode` is the modification of `ObstacleDetectorNode` to implement this improvement. This new node operates with TFs, so at some point, it consult the value of an existing TF. For this reason, in addition to having a `StaticTransformPublisher`, it instantiates a `TransformListener` with its related `Buffer`.

```
include/br2_tf2_detector/ObstacleMonitorNode.hpp

class ObstacleDetectorImprovedNode : public rclcpp::Node
{
...
private:
  ...
  tf2::BufferCore tf_buffer_;
  tf2_ros::TransformListener tf_listener_;
};
```

Let the implementation of this node. In this program, check the two data structures that are related but are not the same:

- `geometry_msgs::msg::TransformStamped` is a message type, and is used to post TFs, and is the returned result of `lookupTransform`.

- `tf2::Transform` is a data type of the TF2 library that allows performing operations.

- `tf2::Stamped<tf2::Transform>` is similar to the previous one, but with a header that indicates a timestamp. It will be necessary to comply with the types in the transformation functions.

- **tf2::fromMsg/tf2::toMsg** are transformation functions that allow transforming from a message type to a TF2 type, and vice versa.

As a general tip, do not use message types inside the node to operate on them. Apply this advice for TFs, images, point clouds, and more data type. Messages are good for communicating nodes but very limited in functionality. If there is a library that offers a native type, use it, as it will be much more useful. Commonly, there are functions to pass from message type to native type. In this case, we use **geometry_msgs::msg::TransformStamped** to send and receive TFs, but we use the TF2 library to operate on them.

Considering the convention established previously, let's see how we can carry out the improvement. Our goal, as we saw before, is to create the TF **odom2object** (object is the detected obstacle). The observation is represented as the transform **laser2object**, so we have to find X in the following equation:

$$odom2object = X * laser2object$$

By deduction from the rules that we stated above, X must be **odom2laser**, which is a TF that can be requested from **lookupTransform**.

```
include/br2_tf2_detector/ObstacleMonitorNode.hpp

    double dist = msg->ranges[msg->ranges.size() / 2];

    if (!std::isinf(dist)) {
      tf2::Transform laser2object;
      laser2object.setOrigin(tf2::Vector3(dist, 0.0, 0.0));
      laser2object.setRotation(tf2::Quaternion(0.0, 0.0, 0.0, 1.0));

      geometry_msgs::msg::TransformStamped odom2laser_msg;
      tf2::Stamped<tf2::Transform> odom2laser;
      try {
        odom2laser_msg = tf_buffer_.lookupTransform(
          "odom", "base_laser_link", msg->header.stamp, rclcpp::Duration(200ms));
        tf2::fromMsg(odom2laser_msg, odom2laser);
      } catch (tf2::TransformException & ex) {
        RCLCPP_WARN(get_logger(), "Obstacle transform not found: %s", ex.what());
        return;
      }

      tf2::Transform odom2object = odom2laser * laser2object;

      geometry_msgs::msg::TransformStamped odom2object_msg;
      odom2object_msg.transform = tf2::toMsg(odom2object);

      odom2object_msg.header.stamp = msg->header.stamp;
      odom2object_msg.header.frame_id = "odom";
      odom2object_msg.child_frame_id = "detected_obstacle";

      tf_broadcaster_->sendTransform(odom2object_msg);
    }
```

- **laser2object** stores the perception to the detected object. It is just a translation in the X-axis corresponding to the distance to the obstacle.

- To get **odom2laser**, we need to query the TF subsystem with `lookupTransform`, transforming the resulting transform message to the needed type to operate with transforms.

- At this point, we have everything to calculate `odom2object`, this is, the obstacle position with respect to the fixed frame `odom`.

- Finally, we compound the output message and publish to the TF subsystem.

It is not necessary to make any changes to the `ObstacleMonitorNode` as `lookupTransform` will calculate the TF `base_footprint` → `obstacle` since the TF system knows the TFs `odom` → `base_footprint` and `odom` → `obstacle`.

4.4.1 Running the Improved Detector

The process of executing the new, improved node is similar to the basic case, with the only difference being our improved node specified in the main program and launcher. Since it is a simple change, we will skip showing it here. Let's follow similar commands to execute it:

```
# Terminal 1: The Tiago simulation
$ ros2 launch br2_tiago sim.launch.py world:=empty
```

```
# Terminal 2: Launch our nodes
$ ros2 launch br2_tf2_detector detector_improved.launch.py
```

```
# Terminal 3: Keyboard teleoperation
$ ros2 run teleop_twist_keyboard teleop_twist_keyboard --ros-args -r
cmd_vel:=/key_vel
```

```
# Terminal 4: RViz2
$ ros2 run rviz2 rviz2 --ros-args -p use_sim_time:=true
```

Add the obstacle in Gazebo so the robot can detect it. Watch the console output and the visual marker in RViz2. Move the robot so the obstacle is not detected, and see how the marker and the output are correct now. The displacement, coded as the transform `odom` → `base_footprint` is used to update the information correctly.

PROPOSED EXERCISES:

1. Make a node that shows every second how much the robot has moved. You can do this by saving $(odom \rightarrow base\_footprint)_t$, and subtracting it from $(odom \rightarrow base\_footprint)_{t+1}$

2. In `ObstacleDetectorNode`, change the arrow's color depending on the distance to the obstacle: green is far, and red is near.

3. In `ObstacleDetectorNode`, show in the terminal the obstacle's position in the odom frame, in the `base_footprint`, and in the `head_2_link`.

## CHAPTER 5

Reactive Behaviors

REACTIVE behaviors tightly couples perception to action without the use of intervening abstract representation. As Brooks demonstrated in his Subsumption Architectures [7], relatively complex behaviors can be created with simple reactive behaviors that are activated or inhibited by higher layers.

We will not discuss the development of sumbsumption architectures in this chapter. By the way, the reader can refer to the Cascade Lifecycle[1] package and rqt_cascade_hfsm[2], which provide some building blocks to build subsumption architectures. The objective of this chapter is to show a couple of reactive behaviors that use different resources to advance the knowledge of ROS 2.

This chapter will first look at a simple local navigation algorithm, Virtual Force Field (VFF), that uses the laser to avoid obstacles. This example will establish some knowledge about visual markers and introduce some test-driven development methodology.

Second, we will see reactive tracking behavior based on information from the camera. We will see how images are processed and how the joints of a robot are controlled. In addition, we will see an advantageous type of node called Lifecycle Node.

5.1 AVOIDING OBSTACLES WITH VFF

This section will show how to implement a simple reactive behavior that makes the Tiago robot move forward, avoiding obstacles using a simple VFF algorithm. This simple algorithm is based on using three 2D vectors to calculate the control speed:

- **Attractive vector**: This vector always points forward since the robot wants to move in a straight line in the absence of obstacles.

- **Repulsive vector**: This vector is calculated from the laser sensor readings. In our basic version, the obstacle closest to the robot produces a repulsion vector, inversely proportional to its distance.

---

[1]https://github.com/fmrico/cascade_lifecycle
[2]https://github.com/fmrico/rqt_cascade_hfsm

DOI: 10.1201/9781003516798-5

- **Result vector**: This vector is the sum of the two previous vectors and will calculate the control speed. Linear speed depends on the resulting vector module, and the angle to turn depends on the resulting vector angle.

Figure 5.1 shows examples of these vectors depending on the position of the obstacles.

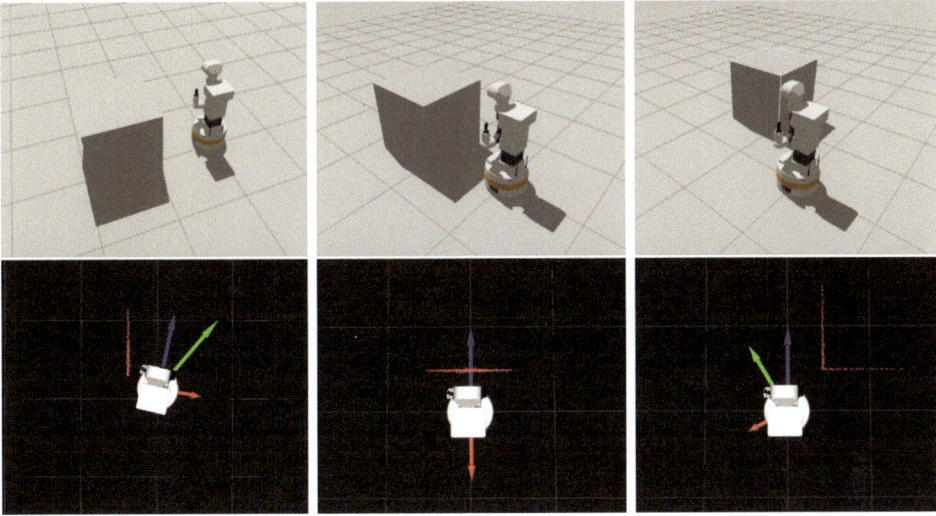

Figure 5.1: Examples of VFF vectors produced by the same obstacle. Blue vector is attractive, red vector is repulsive, and green vector is the resulting vector.

5.1.1   The Computation Graph

First, see what the computational graph of this problem looks like. As shown in Figure 5.2, we have a single node within a process, with the following elements and characteristics:

- The node subscribes to a message topic with the perception information and publishes it to a speed message topic. These will be the main input and outputs. As discussed in the previous chapter, we will use generic names for these topics, which will be remapped at deployment.

- It is crucial to have enough information to determine why a robot behaves in a certain way. ROS 2 offers many debugging tools. Using `/rosout` is a good alternative. It is also handy to use the LEDs equipped by a robot. With an LED that could change color, we could already color-code the robot's state or perception. At a glance, we could have much information about why the robot makes its decisions.

  In this case, in addition to the input and output topics above, we have added the debugging topic `/vff_debug`, that publish Visual Markers to visualize the different vectors of VFF. The color vectors in Figure 5.1 are visual markers published by the node and visualized in RViz2.

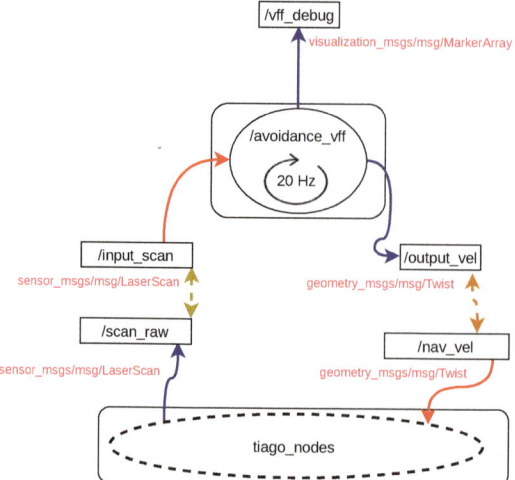

Figure 5.2: Computation Graph for obstacle avoidance.

- In this case, we will choose an iterative execution controlled internally by the node using a timer, to run the control logic at 20 Hz.

5.1.2 Package Structure

See that the organization of the package, in the next box, is already standard in our packages: Each node with its declaration and its different definition in its `.hpp` and its `.cpp`, and the main program that will instantiate it. We have a launch directory with a launcher to easily execute our project. Notice that we have now added a tests directory in which we will have our files with the tests, as we will explain later.

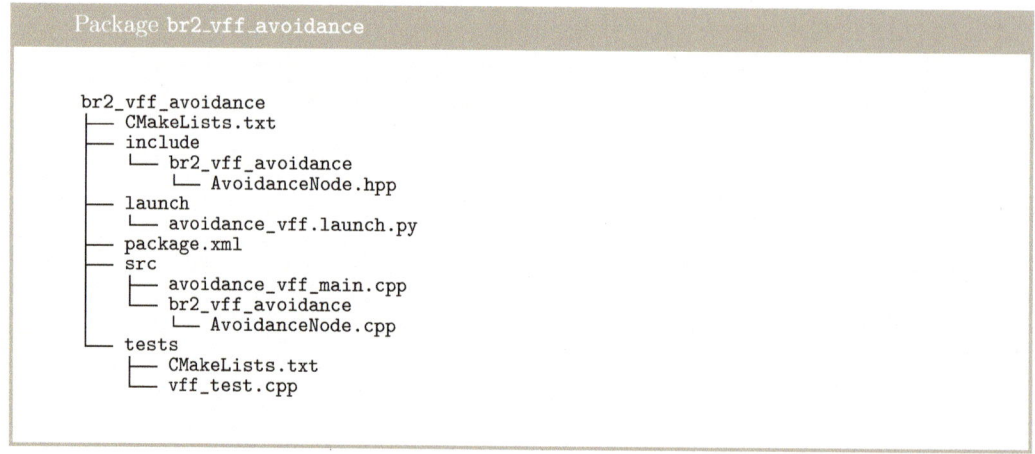

5.1.3 Control Logic

The `AvoidanceNode` implements the VFF algorithm to generate the control commands based on the laser readings. The main elements are similar to the previous examples:

- A subscriber for the laser readings, whose function will be to update the last reading in `last_scan_`.

- A publisher for speeds.

- A `get_vff` function for calculating the three vectors on which the VFF algorithm is based, given a reading from the laser. We declare a new type `VFFVectors` to pack them.

- As this node executes iteratively, we use a timer and use the method `control_cycle` as a callback.

```
include/br2_vff_avoidance/AvoidanceNode.hpp

struct VFFVectors
{
  std::vector<float> attractive;
  std::vector<float> repulsive;
  std::vector<float> result;
};

class AvoidanceNode : public rclcpp::Node
{
public:
  AvoidanceNode();

  void scan_callback(sensor_msgs::msg::LaserScan::UniquePtr msg);
  void control_cycle();

protected:
  VFFVectors get_vff(const sensor_msgs::msg::LaserScan & scan);

private:
  rclcpp::Publisher<geometry_msgs::msg::Twist>::SharedPtr vel_pub_;
  rclcpp::Subscription<sensor_msgs::msg::LaserScan>::SharedPtr scan_sub_;
  rclcpp::TimerBase::SharedPtr timer_;

  sensor_msgs::msg::LaserScan::UniquePtr last_scan_;
};
```

In the control cycle, initially check if the laser has new data. If not, or if this data is old (if we have not received information from the laser in the last second), do not generate control commands. The robot should stop if the robot driver is correctly implemented and does not move when it stops receiving commands. On the contrary (not our case), you should send speeds with all fields to 0 to stop the robot.

Once the resulting vector has been calculated, its transformation at speeds is direct by calculating modulus and angle. It is convenient to control that the speed ranges are in safe ranges with `std::clamp`, as can be seen in the following code:

```
src/br2_vff_avoidance/AvoidanceNode.cpp
```

```cpp
void
AvoidanceNode::scan_callback(sensor_msgs::msg::LaserScan::UniquePtr msg)
{
  last_scan_ = std::move(msg);
}

void
AvoidanceNode::control_cycle()
{
  // Skip cycle if no valid recent scan available
  if (last_scan_ == nullptr || (now() - last_scan_->header.stamp) > 1s) {
    return;
  }

  // Get VFF vectors
  const VFFVectors & vff = get_vff(*last_scan_);

  // Use result vector to calculate output speed
  const auto & v = vff.result;
  double angle = atan2(v[1], v[0]);
  double module = sqrt(v[0] * v[0] + v[1] * v[1]);

  // Create ouput message, controlling speed limits
  geometry_msgs::msg::Twist vel;
  vel.linear.x = std::clamp(module, 0.0, 0.3);   // linear vel to [0.0, 0.3] m/s
  vel.angular.z = std::clamp(angle, -0.5, 0.5);  // rotation vel to [-0.5, 0.5] rad/s

  vel_pub_->publish(vel);
}
```

5.1.4   Calculation of the VFF Vectors

The objective of the function `get_vff` is to obtain the three vectors: attractive, repulsive, and resulting:

```
src/br2_vff_avoidance/AvoidanceNode.cpp
```

```cpp
VFFVectors
AvoidanceNode::get_vff(const sensor_msgs::msg::LaserScan & scan)
{
  // This is the obstacle radius in which an obstacle affects the robot
  const float OBSTACLE_DISTANCE = 1.0;

  // Init vectors
  VFFVectors vff_vector;
  vff_vector.attractive = {OBSTACLE_DISTANCE, 0.0};  // Robot wants to go forward
  vff_vector.repulsive = {0.0, 0.0};
  vff_vector.result = {1.0, 0.0};

  // Get the index of nearest obstacle
  int min_idx = std::min_element(scan.ranges.begin(), scan.ranges.end())
    - scan.ranges.begin();

  // Get the distance to nearest obstacle
  float distance_min = scan.ranges[min_idx];

  // If the obstacle is in the area that affects the robot, calculate repulsive vector
  if (distance_min < OBSTACLE_DISTANCE) {
    float angle = scan.angle_min + scan.angle_increment * min_idx;

    float oposite_angle = angle + M_PI;
    // The module of the vector is inverse to the distance to the obstacle
    float complementary_dist = OBSTACLE_DISTANCE - distance_min;

    // Get cartesian (x, y) components from polar (angle, distance)
    vff_vector.repulsive[0] = cos(oposite_angle) * complementary_dist;
    vff_vector.repulsive[1] = sin(oposite_angle) * complementary_dist;
  }
```

```
src/br2_vff_avoidance/AvoidanceNode.cpp

  // Calculate resulting vector adding attractive and repulsive vectors
  vff_vector.result[0] = (vff_vector.repulsive[0] + vff_vector.attractive[0]);
  vff_vector.result[1] = (vff_vector.repulsive[1] + vff_vector.attractive[1]);

  return vff_vector;
}
```

- The attractive vector will always be $(1, 0)$, since the robot will always try to move forward. Initialize the rest of the vectors assuming there are no nearby obstacles.

- The repulsive vector is calculated from the lower laser reading. By calculating `min_idx` as the index of the vector with a smaller value, we are able to get the distance (the value in the ranges vector) and the angle (from `angle_min`, the `angle_increment` and the `min_idx`).

- The margnitude of the repulsive vector has to be inversely proportional to the distance to the obstacle. Closer obstacles have to generate more repulse than those further.

- The angle of the repulsive vector must be in the opposite direction to the angle of the detected obstacle, so add π to it.

- After calculating the repulsive vector's cartesian coordinates, we add it with the attractive vector to obtain its resultant.

5.1.5 Debugging with Visual Markers

In the previous chapter we used visual markers to visually debug the robot's behavior. The arrows in Figure 5.1 are visual markers generated by `AvoidanceNode` for debugging. The difference is using `visualization_msgs::msg::MarkerArray` instead of `visualization_msgs::msg::Marker`. Basically, a `visualization_msgs::msg::MarkerArray` contains a `std::vector` of `visualization_msgs::msg::Marker` in its field `markers`. Let's see how the message that will be published as debugging information is composed. For details of these messages, check the message definitions, and the reference page[3]:

```
$ ros2 interface show visualization_msgs/msg/MarkerArray

Marker[] markers

$ ros2 interface show visualization_msgs/msg/Marker
```

The `AvoidanceNode` header contains what you need to compose and publish the visual markers. We have a publisher of `visualization_msgs::msg::MarkerArray` and two functions that will help us to compose the vectors. `get_debug_vff` returns the complete message formed by the three arrows that represent the three vectors.

---

[3]http://wiki.ros.org/rviz/DisplayTypes/Marker

To avoid repeating code in this function, `make_marker` creates a marker with the specified color as the input parameter.

```
include/br2_vff_avoidance/AvoidanceNode.hpp

typedef enum {RED, GREEN, BLUE, NUM_COLORS} VFFColor;

class AvoidanceNode : public rclcpp::Node
{
public:
  AvoidanceNode();

protected:
  visualization_msgs::msg::MarkerArray get_debug_vff(const VFFVectors & vff_vectors);
  visualization_msgs::msg::Marker make_marker(
    const std::vector<float> & vector, VFFColor vff_color);

private:
  rclcpp::Publisher<visualization_msgs::msg::MarkerArray>::SharedPtr vff_debug_pub_;
};
```

The markers are published to `control_cycle`, as long as there is a subscriber interested in this information, which, in this case, will be RViz2.

```
void
AvoidanceNode::control_cycle()
{
  // Get VFF vectors
  const VFFVectors & vff = get_vff(*last_scan_);

  // Produce debug information, if any interested
  if (vff_debug_pub_->get_subscription_count() > 0) {
    vff_debug_pub_->publish(get_debug_vff(vff));
  }
}
```

For each of the vectors, create a `visualization_msgs::msg::Marker` with a different color. `base_fooprint` is the frame that is on the ground, in the center of the robot, facing forward. So, the arrow's origin is (0, 0) in this frame, and the arrow's end is what each vector indicates. Each vector must have a different id since a marker will replace another with the same id in RViz2.

```
visualization_msgs::msg::MarkerArray
AvoidanceNode::get_debug_vff(const VFFVectors & vff_vectors)
{
visualization_msgs::msg::MarkerArray marker_array;

marker_array.markers.push_back(make_marker(vff_vectors.attractive, BLUE));
marker_array.markers.push_back(make_marker(vff_vectors.repulsive, RED));
marker_array.markers.push_back(make_marker(vff_vectors.result, GREEN));

return marker_array;
}

visualization_msgs::msg::Marker
AvoidanceNode::make_marker(const std::vector<float> & vector, VFFColor vff_color)
{
  visualization_msgs::msg::Marker marker;

  marker.header.frame_id = "base_footprint";
  marker.header.stamp = now();
  marker.type = visualization_msgs::msg::Marker::ARROW;
  marker.id = visualization_msgs::msg::Marker::ADD;

  geometry_msgs::msg::Point start;
  start.x = 0.0;
```

```
  start.y = 0.0;
  geometry_msgs::msg::Point end;
  start.x = vector[0];
  start.y = vector[1];
  marker.points = {end, start};

  marker.scale.x = 0.05;
  marker.scale.y = 0.1;

  switch (vff_color) {
    case RED:
      marker.id = 0;
      marker.color.r = 1.0;
      break;
    case GREEN:
      marker.id = 1;
      marker.color.g = 1.0;
      break;
    case BLUE:
      marker.id = 2;
      marker.color.b = 1.0;
      break;
  }
  marker.color.a = 1.0;

  return marker;
}
```

5.1.6   Running the `AvoidanceNode`

The main program that runs this node should now be pretty trivial to the reader. Just instantiate the node and call with it to `spin`:

```
src/avoidance_vff_main.cpp

int main(int argc, char * argv[])
{
  rclcpp::init(argc, argv);

  auto avoidance_node = std::make_shared<br2_reactive_behaviors::AvoidanceNode>();
  rclcpp::spin(avoidance_node);
  rclcpp::shutdown();

  return 0;
}
```

To run this node, we must first run the simulator:

```
$ ros2 launch mr2_tiago sim.launch.py
```

Optionally, put again some extra obstacles. Next, execute the node setting remaps and parameters:

```
$ ros2 run br2_vff_avoidance avoidance_vff --ros-args -r input_scan:=/scan_raw -r
output_vel:=/key_vel -p use_sim_time:=true
```

Or using the launcher:

```
$ ros2 launch br2_vff_avoidance avoidance_vff.launch.py
```

If everything goes well, the robot starts to move forward. Use the buttons to move objects in the simulator to put obstacles to the robot. Open RViz2 and add the visualization of topic /vff_debug of type visualization_msgs::msg::MarkerArray, as shown in Figure 5.3. See how the visual information of the node's markers helps us better understand what the robot is doing.

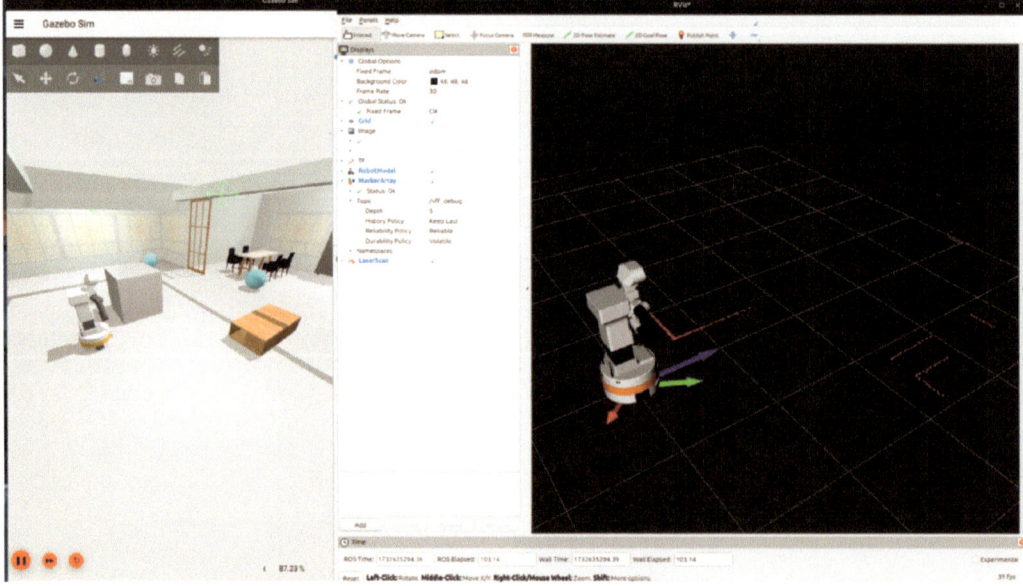

Figure 5.3: Execution of avoidance behavior.

5.1.7 Testing During Development

The code shown in the previous sections may contain calculation errors that can be detected before running it on a real robot and even before running it on a simulator. A very convenient strategy is, taking some (not all) concepts of test-driven development, doing tests simultaneously as the code is developed. This strategy has several advantages:

- Ensure that once a part of the software has been tested, other parts' changes do not negatively affect what has already been developed. The tests are incremental. All tests are always passed, assessing the new functionality and the validity of previously existing code, making development faster.

- The revision task is greatly simplified if the package receives contributions from other developers. Activating a CI (Continous Integration) system in your repository allows that each contribution has to compile correctly and pass all the tests, both functional and stylish. In this way, the reviewer focuses on verifying that the code does its job correctly.

- Many quality assurance procedures require the software to be tested. Saying "I will do the tests when I finish" is a fallacy: *You will not do them*, or it will be a tedious process that will not help you, so they are likely to be incomplete and ineffective.

ROS 2 provides many testing tools that we can use easily. Let's start with the unit tests. ROS 2 uses GoogleTest[4] to test C++ code. In order to use tests in the package, include some packages in the `package.xml`:

```
package.xml

<test_depend>ament_lint_auto</test_depend>
<test_depend>ament_lint_common</test_depend>
<test_depend>ament_cmake_gtest</test_depend>
```

The `<test_depend>` tag contains those dependencies only needed to test the package. It is possible to compile a workspace, in this case only the package, excluding the tests, so these packages will not be taken into account in the dependencies:

```
$ colcon build --symlink-install --packages-select br2_vff_avoidance
--cmake-args -DBUILD_TESTING=off
```

As shown in the package structure, there is a `tests` directory with a C++ file (`vff_test.cpp`) that contains tests. To compile it, these sentences should be in `CMakeLists.txt`:

---

[4]https://github.com/google/googletest

```
CMakeLists.txt

if(BUILD_TESTING)
  find_package(ament_lint_auto REQUIRED)
  ament_lint_auto_find_test_dependencies()

  set(ament_cmake_cpplint_FOUND TRUE)
  ament_lint_auto_find_test_dependencies()

  find_package(ament_cmake_gtest REQUIRED)
  add_subdirectory(tests)
endif()
```

```
tests/CMakeLists.txt

ament_add_gtest(vff_test vff_test.cpp)
ament_target_dependencies(vff_test ${dependencies})
target_link_libraries(vff_test ${PROJECT_NAME})
```

Once introduced the testing infrastructure in a package, see how to do unit tests. While developing the method `AvoidanceNode::get_vff` it is possible to check that it works correctly. Just create several synthetic `sensor_msgs::msg::LaserScan` messages and then check that this function returns correct values in all cases. In this file, it has been developed eight different cases. Let's see some of them:

```
tests/vff_test.cpp

sensor_msgs::msg::LaserScan get_scan_test_1(rclcpp::Time ts)
{
  sensor_msgs::msg::LaserScan ret;
  ret.header.stamp = ts;
  ret.angle_min = -M_PI;
  ret.angle_max = M_PI;
  ret.angle_increment = 2.0 * M_PI / 16.0;
  ret.ranges = std::vector<float>(16, std::numeric_limits<float>::infinity());

  return ret;
}

sensor_msgs::msg::LaserScan get_scan_test_5(rclcpp::Time ts)
{
  sensor_msgs::msg::LaserScan ret;
  ret.header.stamp = ts;
  ret.angle_min = -M_PI;
  ret.angle_max = M_PI;
  ret.angle_increment = 2.0 * M_PI / 16.0;
  ret.ranges = std::vector<float>(16, 5.0);
  ret.ranges[10] = 0.3;

  return ret;
}
```

Each function returns a `sensor_msgs::msg::LaserScan` message as if it had been generated by a laser with 16 different values, regularly distributed in the range $[-\pi, \pi]$. In `get_scan_test_1` it simulates the case that no obstacles are detected in any case. At `get_scan_test_5` it simulates that there is an obstacle at position 10, which corresponds to angle $-\pi + 10 * \frac{2\pi}{16} = 0.785$.

In order to access the method to be tested, since it is not public, it is convenient to make it `protected` and implement a class to access these functions:

```
tests/vff_test.cpp

class AvoidanceNodeTest : public br2_vff_avoidance::AvoidanceNode
{
public:
  br2_vff_avoidance::VFFVectors
  get_vff_test(const sensor_msgs::msg::LaserScan & scan)
  {
    return get_vff(scan);
  }

  visualization_msgs::msg::MarkerArray
  get_debug_vff_test(const br2_vff_avoidance::VFFVectors & vff_vectors)
  {
    return get_debug_vff(vff_vectors);
  }
};
```

It is possible to have all the needed tests in the same file. Each of them is defined using the macro TEST(id, sub_id), and inside, as if it were a function, write a program whose objective is to test the functionality of the code. In the case of get_vff, these are the **unitary tests**:

```
tests/vff_test.cpp

TEST(vff_tests, get_vff)
{
  auto node_avoidance = AvoidanceNodeTest();
  rclcpp::Time ts = node_avoidance.now();

  auto res1 = node_avoidance.get_vff_test(get_scan_test_1(ts));
  ASSERT_EQ(res1.attractive, std::vector<float>({1.0f, 0.0f}));
  ASSERT_EQ(res1.repulsive, std::vector<float>({0.0f, 0.0f}));
  ASSERT_EQ(res1.result, std::vector<float>({1.0f, 0.0f}));

  auto res2 = node_avoidance.get_vff_test(get_scan_test_2(ts));
  ASSERT_EQ(res2.attractive, std::vector<float>({1.0f, 0.0f}));
  ASSERT_NEAR(res2.repulsive[0], 1.0f, 0.00001f);
  ASSERT_NEAR(res2.repulsive[1], 0.0f, 0.00001f);
  ASSERT_NEAR(res2.result[0], 2.0f, 0.00001f);
  ASSERT_NEAR(res2.result[1], 0.0f, 0.00001f);

  auto res5 = node_avoidance.get_vff_test(get_scan_test_5(ts));
  ASSERT_EQ(res5.attractive, std::vector<float>({1.0f, 0.0f}));
  ASSERT_LT(res5.repulsive[0], 0.0f);
  ASSERT_LT(res5.repulsive[1], 0.0f);
  ASSERT_GT(atan2(res5.repulsive[1], res5.repulsive[0]), -M_PI);
  ASSERT_LT(atan2(res5.repulsive[1], res5.repulsive[0]), -M_PI_2);
  ASSERT_LT(atan2(res5.result[1], res5.result[0]), 0.0);
  ASSERT_GT(atan2(res5.result[1], res5.result[0]), -M_PI_2);
}

int main(int argc, char ** argv)
{
  rclcpp::init(argc, argv);
  testing::InitGoogleTest(&argc, argv);
  return RUN_ALL_TESTS();
}
```

The ASSERT_* macros check the expected values based on the input. ASSERT_EQ verifies that the two values are equal. When comparing floats, it is preferable to use ASSERT_NEAR, which checks that two values are equal with a specified margin in its third parameter. ASSERT_LT verifies that the first value is "Less Than" the second. ASSERT_GT verifies that the first value is "Greater Than" the second, and so on.

For example, case 5 (obstacle at angle 0.785) verifies that the coordinates of the repulsive vector are negative, both its angle is in the range $[-\pi, -\frac{\pi}{2}]$ (it is a vector opposite to angle 0.785) and that the resulting vector is in the range $[0, -\frac{\pi}{2}]$. If this is true, the algorithm is correct. Do these checks for each reading, with its expected values, and pay attention to extreme and unexpected cases, such as test 1.

It is also possible to do **integration tests**. Since nodes are objects, instantiate them and simulate their operation. For example, test the speeds published by AvoidanceNode when receiving the test messages. Let's see how to do it:

```
tests/vff_test.cpp

TEST(vff_tests, ouput_vels)
{
  auto node_avoidance = std::make_shared<AvoidanceNodeTest>();

  // Create a testing node with a scan publisher and a speed subscriber
  auto test_node = rclcpp::Node::make_shared("test_node");
  auto scan_pub = test_node->create_publisher<sensor_msgs::msg::LaserScan>(
    "input_scan", 100);

  geometry_msgs::msg::Twist last_vel;
  auto vel_sub = test_node->create_subscription<geometry_msgs::msg::Twist>(
    "output_vel", 1, [&last_vel] (geometry_msgs::msg::Twist::SharedPtr msg) {
      last_vel = *msg;
    });

  ASSERT_EQ(vel_sub->get_publisher_count(), 1);
  ASSERT_EQ(scan_pub->get_subscription_count(), 1);

  rclcpp::Rate rate(30);
  rclcpp::executors::SingleThreadedExecutor executor;
  executor.add_node(node_avoidance);
  executor.add_node(test_node);

  // Test for scan test #1
  auto start = node_avoidance->now();
  while (rclcpp::ok() && (node_avoidance->now() - start) < 1s) {
    scan_pub->publish(get_scan_test_1(node_avoidance->now()));
    executor.spin_some();
    rate.sleep();
  }
  ASSERT_NEAR(last_vel.linear.x, 0.3f, 0.0001f);
  ASSERT_NEAR(last_vel.angular.z, 0.0f, 0.0001f);

  // Test for scan test #2
}
```

1. Create an AvoidanceNodeTest (AvoidanceNode is also possible) node to test it.

2. Make a generic node called test_node to create a laser scan publisher and a speed subscriber.

3. When creating the speed subscriber, a lambda function has especified as a callback. This lambda function accesses the last_vel variable to update it with the last message received at the topic output_vel.

4. Create an executor and add both nodes to it to execute them.

5. During a second post at 30 Hz on input_scan a sensor reading corresponding to the synthetic readings.

6. In the end, verify that the published speeds are correct.

To run just these gtest tests, do it by running the binary that is in the tests directory of the package, in the build directory:

```
$ cd ~/bookros2_ws

$ build/br2_vff_avoidance/tests/vff_test

[==========] Running 2 tests from 1 test case.
[----------] Global test environment set-up.
[----------] 2 tests from vff_tests
[ RUN      ] vff_tests.get_vff
[       OK ] vff_tests.get_vff (18 ms)
[ RUN      ] vff_tests.ouput_vels
[       OK ] vff_tests.ouput_vels (10152 ms)
[----------] 2 tests from vff_tests (10170 ms total)

[----------] Global test environment tear-down
[==========] 2 tests from 1 test case ran. (10170 ms total)
[ PASSED   ] 2 tests.
```

To run all the tests for this package, even the style ones, use `colcon`:

```
$ colcon test --packages-select br2_vff_avoidance
```

If the test has finished with failures, go to check what has failed to the directory `log/latest_test/br2_vff_avoidance/stdout_stderr.log`. At the end of the file, there is a summary of the failed tests. For example, this message at the end indicates that tests 3, 4, 5, and 7 failed (errors were intentionally added for this explanation):

```
log/latest_test/br2_vff_avoidance/stdout_stderr.log

56% tests passed, tests failed out of 9

Label Time Summary:
copyright     =    0.37 sec*proc (1 test)
cppcheck      =    0.44 sec*proc (1 test)
cpplint       =    0.45 sec*proc (1 test)
flake8        =    0.53 sec*proc (1 test)
gtest         =   10.22 sec*proc (1 test)
lint_cmake    =    0.34 sec*proc (1 test)
linter        =    3.88 sec*proc (8 tests)
pep257        =    0.38 sec*proc (1 test)
uncrustify    =    0.38 sec*proc (1 test)
xmllint       =    0.99 sec*proc (1 test)

Total Test time (real) =   14.11 sec

The following tests FAILED:
        [   3 - cpplint (Failed)]
        [   4 - flake8 (Failed)]
        [   5 - lint_cmake (Failed)]
        [   7 - uncrustify (Failed)]
Errors while running CTest
```

Each line in this file begins with the section number corresponding to a test. Go, for example, to sections 3, 4, and 7 to see some of these errors:

```
log/latest_test/br2_vff_avoidance/stdout_stderr.log

3: br2_vff_avoidance/tests/vff_test.cpp:215:  Add #include <memory> for
   make_shared<>  [build/include_what_you_use] [4]
3: br2_vff_avoidance/include/br2_vff_avoidance/AvoidanceNode.hpp:15:  #ifndef header
   guard has wrong style, please use: BR2_VFF_AVOIDANCE__AVOIDANCENODE_HPP_
   [build/header_guard] [5]

4: ./launch/avoidance_vff.launch.py:34:3: E111 indentation is not a multiple of four
4:   ld.add_action(vff_avoidance_cmd)
4:   ^

7: --- src/br2_vff_avoidance/AvoidanceNode.cpp
7: +++ src/br2_vff_avoidance/AvoidanceNode.cpp.uncrustify
7: @@ -100,2 +100 @@
7: -  if (distance_min < OBSTACLE_DISTANCE)
7: -  {
7: +  if (distance_min < OBSTACLE_DISTANCE) {
7: @@ -109 +108 @@
7: -    vff_vector.repulsive[0] = cos(oposite_angle)*complementary_dist;
7: +    vff_vector.repulsive[0] = cos(oposite_angle) * complementary_dist;
7:
7: Code style divergence in file 'tests/vff_test.cpp':
```

- The errors in Section 3 correspond to cpplint, a C++ linter. The first error indicates that a header must be added since there are functions that are declared in it. In the second, it indicates that the style of the header guard in AvoidanceNode.hpp is incorrect, indicating which one should be used.

- The errors in section 4 correspond to flake8, a Python linter. This error indicates that the launcher file uses an incorrect indentation since it should be space, multiples of 4.

- The errors labeled with 7 correspond to uncrustify, another C++ linter. In a format similar to the output of the diff command, it tells the difference between the code that is written and the one that should be in good style. In this case, it indicates that the start of an if block on line 100 of AvoidanceNode.cpp should be on the same line as if. The second error indicates that there should be spaces on both sides of an operator.

The first time facing solving style problems, it can seem like a daunting task without much meaning. You would wonder why the style that it indicates is better than yours. Indeed you have been using this style for years, and you are very proud of how your source code looks like. You will not understand why you have to use two spaces in C++ to indent, and not the tab, for example, or why open the blocks in the same line of a while if you always opened in the next line.

The first reason is that it indicates a good style. Cpplint, for example, uses the Google C++ Style Guide[5], which is a widely accepted style guide adopted by most software development companies.

The second is because you have to follow this style if you want to contribute to a ROS 2 project or repository. Rarely a repository that accepts contributions does not have a continuous integration system that passes these tests. Imagine that you are the one who maintains a project. You'll want all of your code to have a consistent

---

[5]https://google.github.io/styleguide/cppguide.html

style. It would be a nightmare to make your own style guide or discuss with each contributor at every pull request style issues rather to focus on their contribution. The worst discussion I can recall with a colleague was using tabs against spaces. It is a discussion that will have no solution because it is like talking about religions. Using a standard solves these problems.

Furthermore, the last reason is that it will make you a better programmer. Most of the style rules have a practical reason. Over time, you will automatically apply the style you have corrected so many times when passing the tests, and your code will have a good style as you write it.

5.2 TRACKING OBJECTS

This section analyzes a project that contains other reactive behavior. In this case, the behavior tracks the objects that match a specific color with the robot's head.

There are several new concepts that are introduced in this project:

- **Image analysis**: So far, we have used a relatively simple sensor. Images provide more complex perceptual information from which a lot of information can be extracted. Remember that there is an essential part of Artificial Intelligence that deals with Artificial Vision, and it is one of the primary sensors in robots. We will show how to process these images with OpenCV, the reference library in this area.

- **Control at joint level**: In the previous projects, the commands were speeds sent to the robot. In this case, we will see how to command positions directly to the joints of the robot's neck.

- **Lifecycle Nodes**: ROS 2 provides a particular type of Node called Lifecycle Node. This node is very useful to control the life cycle, including its startup, activation, and deactivation.

5.2.1 Perception and Actuation Models

This project uses the images from the robot's camera as a source of information. Whenever a node transmits an (non-compressed) image in ROS 2, it uses the same type of message: `sensor_msgs/msg/Image`. All the drivers of all cameras supported in ROS 2 use it. See what the message format is:

```
$ ros2 interface show sensor_msgs/msg/image

# This message contains an uncompressed image
# This message contains an uncompressed

std_msgs/Header header   # Header timestamp should be acquisition time of image
                         # Header frame_id should be optical frame of camera
                         # origin of frame should be optical center of camera
                         # +x should point to the right in the image
                         # +y should point down in the image
                         # +z should point into to plane of the image
                         # If the frame_id and the frame_id of the CameraInfo
                         # message associated with the image conflict
                         # the behavior is undefined

uint32 height            # image height, that is, number of rows
uint32 width             # image width, that is, number of columns

# The legal values for encoding are in file src/image_encodings.cpp
# If you want to standardize a new string format, join
# ros-users@lists.ros.org and send an email proposing a new encoding.

string encoding    # Encoding of pixels -- channel meaning, ordering, size
                   # from the list in include/sensor_msgs/image_encodings.hpp

uint8 is_bigendian # is this data bigendian?
uint32 step        # Full row length in bytes
uint8[] data       # actual matrix data, size is (step * rows)
```

Camera drivers often publish (only once, in transient local QoS) information about camera parameters as a `sensor_msgs/msg/CameraInfo` message, which includes intrinsic and distortion parameters, projection matrix, and more. With this information, we can work with stereo images, for example, or we can combine this information with a depth image to reconstruct the 3D scene. The process of calibrating a camera[6] has to do with calculating the values that are published in this message. A good exercise is reading this message format, although it is not used in this chapter.

Although it is possible to use a simple `sensor_msgs/msg/Image` publisher or subscriber, it is usual when working with images using different transport strategies (compression, streaming codecs ...) using specific publishers/subscribers. The developer uses them and ignores how the images are transported—he just sees an `sensor_msgs/msg/Image`. Check available transport plugins typing:

---

[6]http://wiki.ros.org/image_pipeline

```
$ ros2 run image_transport list_transports

Declared transports:
image_transport/compressed
image_transport/compressedDepth
image_transport/raw
image_transport/theora

Details:
----------

...
```

Run the simulated Tiago and check the topics to see that there is more than one topic for 2D images:

```
$ ros2 topic list

/head_front_camera/image_raw/compressed
/head_front_camera/image_raw/compressedDepth
/head_front_camera/image_raw/theora
/head_front_camera/rgb/camera_info
/head_front_camera/rgb/image_raw
```

The developer has not created all these topics one by one but has used an `image_transport::Publisher` that has generated all these topics taking into account the available transport plugins. In the same way, to obtain the images, it is convenient to use an `image_transport::Subscriber`, as we will see below. Using compressed images may be good if the image is big or the network reliability is not the best. The trade-off is a bit more CPU load on the source and destination.

The image message format is for transporting images, not for processing them. It is not common to work directly with images as raw byte sequences. The usual way is to use some image processing library, and the most widely used is OpenCV[7]. OpenCV provides several hundreds of computer vision algorithms.

The main data type that OpenCV uses to work with images is `cv::Mat`. ROS 2 provides tools to transform `sensor_msgs/msg/Image` into `cv::Mat`, and vice versa:

```cpp
void image_callback(const sensor_msgs::msg::Image::ConstSharedPtr & msg)
{
  cv_bridge::CvImagePtr cv_ptr;
  cv_ptr = cv_bridge::toCvCopy(msg, sensor_msgs::image_encodings::BGR8);
  cv::Mat & image_src = cv_ptr->image;

  sensor_msgs::msg::Image image_out = *cv_ptr->toImageMsg();
}
```

In the perception model of our project, the segmentation of an image will be done by color. It is convenient to work in HSV[8], instead of RGB, which is the encoding in which we receive the messages. HSV encoding represents a pixel in color with three components: Hue, Saturation, and Value. Working in HSV allows us to establish color ranges more robustly to lighting changes since this is what the V component is mainly

---

[7]https://docs.opencv.org/5.x/d1/dfb/intro.html
[8]https://en.wikipedia.org/wiki/HSL_and_HSV

responsible for, and if the range is wider, we can continue to detect the same color even if the illumination changes.

The following code transforms a `cv::mat` to HSV and calculates an image mask with the pixels that match the color of the furniture in the default simulated world of Tiago in Gazebo, as shown in Figure 5.4:

```
cv::Mat img_hsv;
cv::cvtColor(cv_ptr->image, img_hsv, cv::COLOR_BGR2HSV);

cv::Mat1b filtered;
cv::inRange(img_hsv, cv::Scalar(15, 50, 20), cv::Scalar(20, 200, 200), filtered);
```

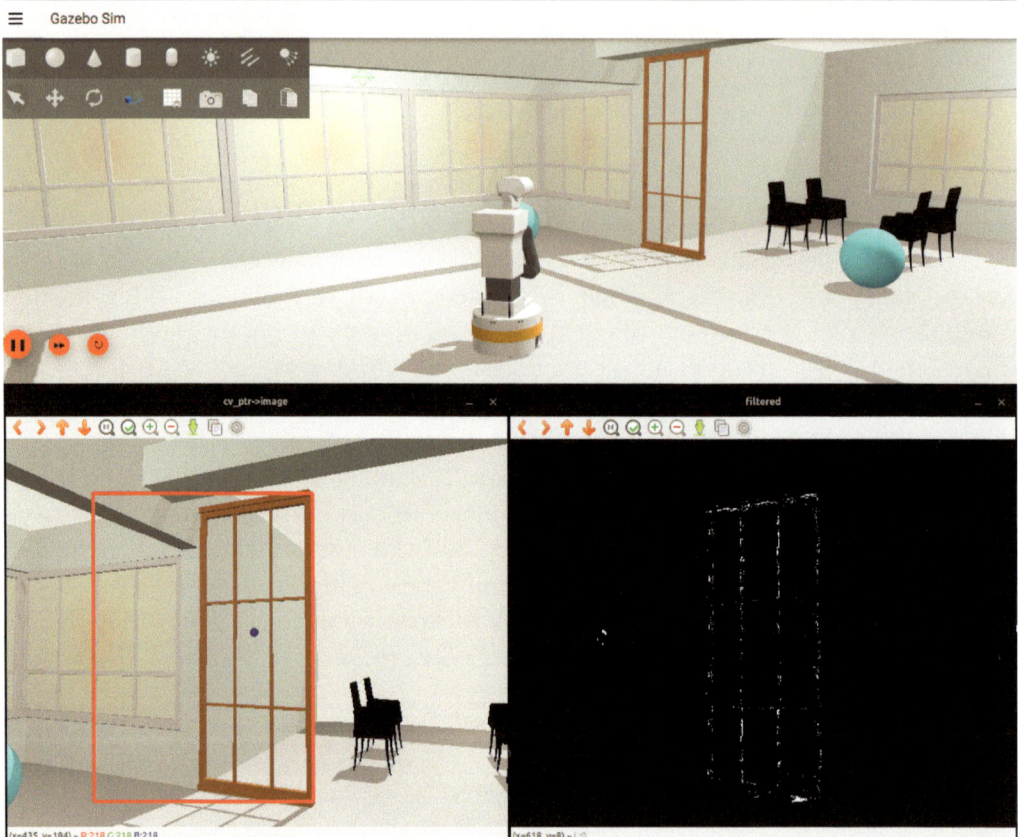

Figure 5.4: Object detection by color using an HSV range filter.

Finally, the output of the processing of an image in this project is a message of type `vision_msgs/msg/Detection2D` (examine the fields in this message for yourself), from which we use its `header`, `bbox`, and `source_img field`. It is not required to use all the fields. The original image is included to have the image's dimensions where the detection is made, whose importance will be shown in the following.

The **action model** is a position control of the robot head. The robot has two joints that control the position the camera is pointing at: `head_1_joint` for horizontal control (pan) and `head_2_joint` for vertical control (tilt).

In ROS 2, the control of the joints is done by the framework **ros2_control**[9]. The developers of the simulated Tiago robot have used a trajectory controller (joint_trajectory_controller) for the two joints of the robot's neck. Through two topics (as shown in Figure 5.5), it allows reading the state of the joints and sending commands in the form of a set of waypoints (Figure 5.6) to be reached at specific time instants. Waypoints consist of positions and optionally velocities, accelerations, and effort, as well as a time from start to be applied.

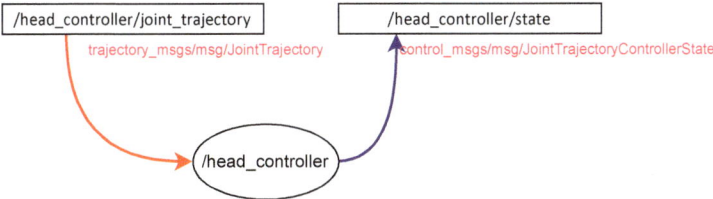

Figure 5.5: Head controller topics.

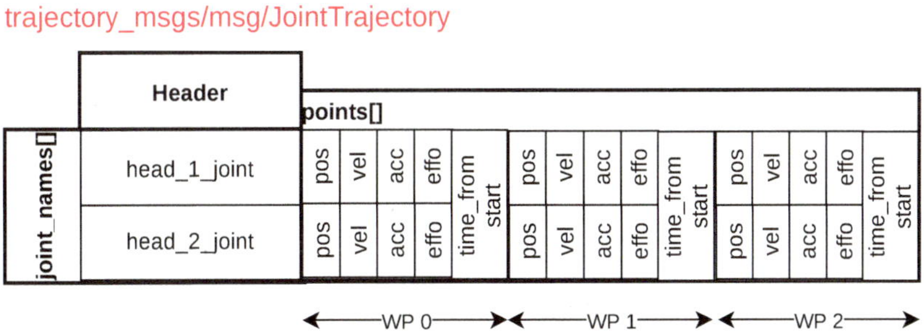

trajectory_msgs/msg/JointTrajectory

Figure 5.6: `trajectory_msgs/msg/JointTrajectory` message format.

Obtaining the 3D position of the object to the robot and calculating the position of the neck joints to center it in the image would probably be an adequate solution in a real active vision system, but quite complex at the moment. We will simply implement a control in the image domain.

The node that controls the robot's neck receives two values (called error) that indicate the difference between the current position and the desired position for pan and tilt, respectively. If a value is 0, it indicates that it is in the desired position. If it is less than 0, the joint has to move in one direction, and greater than zero has to move in the other direction. The range of values is $[-1, +1]$ for each joint, as shown in Figure 5.7. As this node performs iterative control and neck movements can be very fast, a PID controller will control the position to which each joint is commanded to correct its speed.

---

[9]http://control.ros.org/index.html

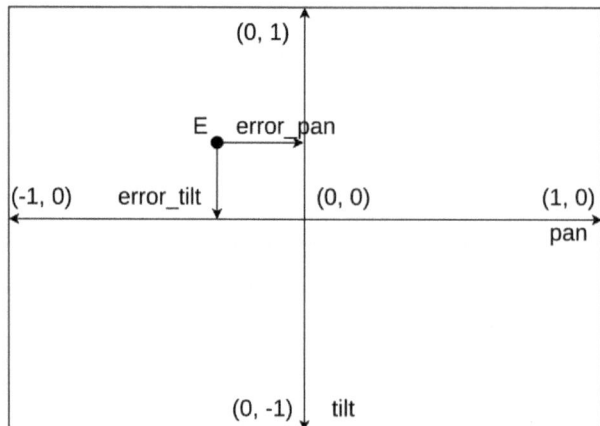

Figure 5.7: Diagram for pan/tilt control. **E** indicates the desired position. `error_*` indicates the difference between the current position and the desired pan/tilt position.

5.2.2 Computation Graph

The Computation Graph of this project (Figure 5.8) shows how this problem is divided into three nodes within the same process. The reason is that each node (`ObjectDetector` and `HeadController`) can be executed separately, and be reused in other problems (we will do it in next chapters). Each one has been designed in this way to be reusable, with inputs and outputs that try to be generic, not strongly coupled to this problem.

In this Computation Graph, the `HeadController` has been represented differently from the rest of the nodes. This node will be implemented as **LifeCycle Node**, which we will explain in the Section 5.2.3. For now, we will say that it is like a standard node but that it can be activated and deactivated during its operation.

The `HeadController` receives a pan/tilt speed, each in the range $[-1, 1]$. Note that since there is no standard ROS 2 message that fits our problem (we could have used `geometry_msgs/msg/Pose2D`, ignoring the field theta), we have created a custom `br2_tracking_msgs/msg/PanTiltCommand` message containing the needed information. We will see below how we have done to create our custom message.

The `ObjectDetector` publishes, for each image, the result of the detection of the furniture in the image. It will return the coordinate, in pixels, of the detection, as well as the bounding box of the object.

The output of the `ObjectDetector` does not completely match the input of the `HeadController`. `ObjectDetector` publishes its output in pixels. In this case, the image resolution is 640×480 so its range is $[0, 640]$ for the horizontal X component and $[0, 480]$ for the vertical Y component. Therefore, we create a node, `tracker`, with a straightforward task, which is to adapt the output of the `ObjectDetector` to the input of the `HeadController`, to make a control in the image, moving the head so that the detected object is always in the center of the image.

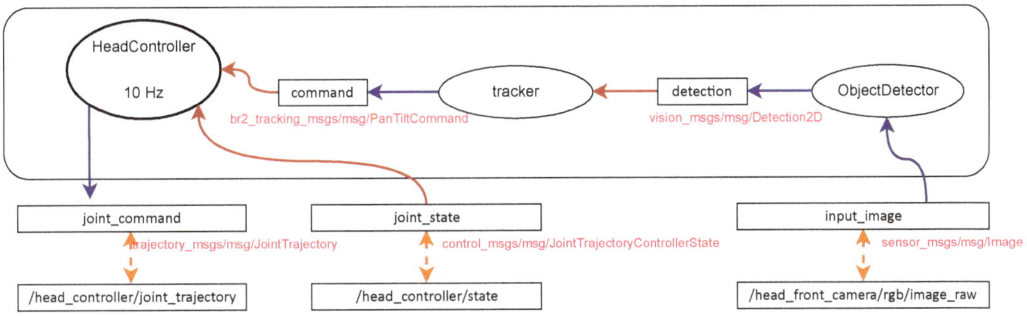

Figure 5.8: Computation Graph for Object Tracking project.

5.2.3   Lifecycle Nodes

So far, we have seen that the nodes in ROS 2 are objects of class `Node` that inherit methods that allow us to communicate with other nodes or obtain information. In ROS 2, there is a type of node, the **LifeCycleNode**, whose lifetime is defined using states and the transitions between them:

- When a LifeCycleNode is created, it is in *Unconfigured* state, and it must trigger the *configure* transition to enter the *Inactive* state.

- A LifeCycleNode is working when it is in the *Active* state, from which it can transition from the *Inactive* state through the *activate* transition. It is also possible to transition from the *Active* to *Inactive* state through the *deactivate* transition.

- The necessary tasks and checks can be performed at each transition. Even a transition can fail and not transit if the conditions specified in the code of its transition are not met.

- In case of error, the node can go to *Finalized* state.

- When a node has completed its task, it can transition to *Finalized*.

See a diagram of these states and transitions in Figure 5.9.
Lifecycle nodes provide a node execution model that allows:

- Make them predictable. For example, in ROS 2, the parameters should be read only in the *configuring* transition.

- When there are multiple Nodes, we can coordinate their startup. We can define that specific nodes are not activated until they are configured. We also can specify some orders in the startup.

- Programmatically, it allows having another option beyond the constructor to start its components. Remember that in C++, a Node is not completely built until its constructor has finished. This usually brings problems if we require a `shared_ptr` to `this`.

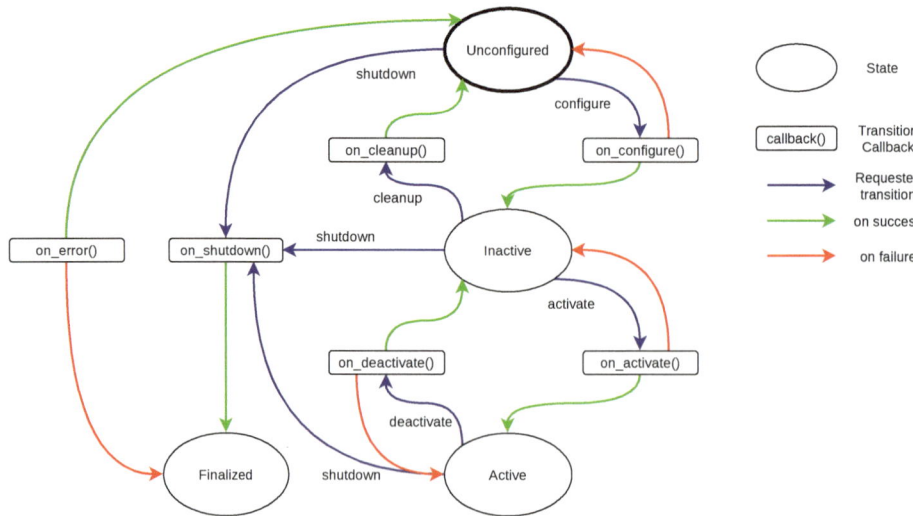

Figure 5.9: Diagram of states and transitions in Lifecycle Nodes.

An example could be a sensor driver. If the physical device cannot be accessed, it cannot transit to the *Inactive* state. In addition, all the initial setup time of the device would be set to this state so that its activation would be immediate. Another example is the startup of a robot driver. It would not boot until all its sensor/actuator nodes are in the *Active* state.

5.2.4 Creating Custom Messages

We have previously specified that the input of node `HeadController` is of type `br2_tracking_msgs/msg/PanTiltCommand` because there was no type of message that conformed to what we needed. One golden rule in ROS 2 is *not to create a message if there is already a standard available*, as we can benefit from available tools for this message. In this case, no standard will serve our purposes. In addition, it is the perfect excuse to show how to create custom messages.

First of all, *when creating new messages* (new interfaces, in general), even in the context of a specific package, it is highly recommended that you *make a separate package*, ending in _msgs. Tools may exist in the future that needs to receive messages of this new type, but we do not necessarily have to depend on the packages for which they were created.

Next we show the structure of package `br2_tracking_msgs` that contains only the definition of message `br2_tracking_msgs/msg/PanTiltCommand`:

```
br2_tracking_msgs/
├── CMakeLists.txt
├── msg
│   └── PanTiltCommand.msg
└── package.xml
```

Packages that contain interfaces have, besides a `package.xml` and a `CMakeLists.txt`, a directory for each type of interface (message, service, or action) that is being defined. In our case, it is a message, so we will have a `msg` directory that contains a `.msg` file for each new message to define. Let's see the definition of the `PanTiltCommand` message:

```
msg/PanTiltCommand.msg

float64 pan
float64 tilt
```

The important part in the `CMakeLists.txt` is the `rosidl_generate_interfaces` statement, in which we specify where the interface definitions are:

```
CMakeLists.txt

find_package(ament_cmake REQUIRED)
find_package(builtin_interfaces REQUIRED)
find_package(rosidl_default_generators REQUIRED)

rosidl_generate_interfaces(${PROJECT_NAME}
  "msg/PanTiltCommand.msg"
  DEPENDENCIES builtin_interfaces
)

ament_export_dependencies(rosidl_default_runtime)
ament_package()
```

5.2.5   Tracking Implementation

The structure of the `br2_tracking` package, shown in the following, follows the guidelines already recommended above.

```
Package br2_vff_avoidance


        br2_tracking
        ├── CMakeLists.txt
        ├── config
        │   └── detector.yaml
        ├── include
        │   └── br2_tracking
        │       ├── HeadController.hpp
        │       ├── ObjectDetector.hpp
        │       └── PIDController.hpp
        ├── launch
        │   └── tracking.launch.py
        ├── package.xml
        ├── src
        │   ├── br2_tracking
        │   │   ├── HeadController.cpp
        │   │   ├── ObjectDetector.cpp
        │   │   └── PIDController.cpp
        │   └── object_tracker_main.cpp
        └── tests
            ├── CMakeLists.txt
            └── pid_test.cpp
```

- The `HeadController` and `ObjectDetector` nodes will be compiled as libraries independently of the main program `object_tracker_main.cpp`. The latter will

be in `src`, while the nodes will have their headers in `include/br2_tracking`
and their definitions in `src/br2_tracking`.

- The library also includes a class to use PID controllers, used in
  `HeadController`.

- A launcher will launch the executable with the necessary parameters and
  remaps.

- There is a `config` directory that contains a YAML file with the HSV range
  that `ObjectDetector` will use to detect in the image the furniture of Tiago's
  default stage in Gazebo.

- The `tests` directory includes tests for the PID controller.

The reader will have noticed that there is no file for a `tracker` node in this structure. This node, being so simple, has been implemented in `object_tracker_main.cpp`
as follows:

```
src/object_tracker_main.cpp

auto node_detector = std::make_shared<br2_tracking::ObjectDetector>();
auto node_head_controller = std::make_shared<br2_tracking::HeadController>();
auto node_tracker = rclcpp::Node::make_shared("tracker");

auto command_pub = node_tracker->create_publisher<br2_tracking_msgs::msg::PanTiltCommand>(
  "/command", 100);
auto detection_sub = node_tracker->create_subscription<vision_msgs::msg::Detection2D>(
  "/detection", rclcpp::SensorDataQoS(),
  [command_pub](vision_msgs::msg::Detection2D::SharedPtr msg) {
    br2_tracking_msgs::msg::PanTiltCommand command;
    command.pan = (msg->bbox.center.x / msg->source_img.width) * 2.0 - 1.0;
    command.tilt = (msg->bbox.center.y / msg->source_img.height) * 2.0 - 1.0;
    command_pub->publish(command);
});

rclcpp::executors::SingleThreadedExecutor executor;
executor.add_node(node_detector);
executor.add_node(node_head_controller->get_node_base_interface());
executor.add_node(node_tracker);
```

`node_tracker` is a generic ROS 2 node, from which we construct a publisher to the
`/command` topic, and a subscriber to the `/detection`. We have specified the subscriber
callback as a lambda function that takes from the input message the position in pixels
of the detected object, together with the size of the image, and generates the inputs
for node `HeadController`, following the scheme already shown in Figure 5.7.

Notice that when adding the node `node_head_controller` to `executor`, we have
used the `get_node_base_interface` method. This is because it is a LifeCycleNode,
as we introduced earlier, and `add_node` does not yet support adding this type of node
directly. Fortunately, we can do it through a basic interface supported by LifeCycleNode and regular nodes using this method.

The `ObjectDetector` will be a `rclcpp::Node`, with an image subscriber (using
`image_transport`) and a 2D detection message publisher. There are two member
variables that will be used in the detection process.

```
include/br2_tracking/ObjectDetector.hpp

class ObjectDetector : public rclcpp::Node
{
public:
  ObjectDetector();

  void image_callback(const sensor_msgs::msg::Image::ConstSharedPtr & msg);

private:
  image_transport::Subscriber image_sub_;
  rclcpp::Publisher<vision_msgs::msg::Detection2D>::SharedPtr detection_pub_;

  // HSV ranges for detection [h - H] [s - S] [v - V]
  std::vector<double> hsv_filter_ranges_ {0, 180, 0, 255, 0, 255};
  bool debug_ {true};
};
```

These variables, with a default value, will be initialized using parameters. They are the HSV color ranges and a variable that, by default, causes a window to be displayed with the detection result for debugging purposes.

```
src/br2_tracking/ObjectDetector.cpp

ObjectDetector::ObjectDetector()
: Node("object_detector")
{
  declare_parameter("hsv_ranges", hsv_filter_ranges_);
  declare_parameter("debug", debug_);

  get_parameter("hsv_ranges", hsv_filter_ranges_);
  get_parameter("debug", debug_);
}
```

When executing the program with all the nodes, a parameter file in the config directory will be specified to set the color filter.

```
config/detector.yaml

/object_detector:
  ros__parameters:
    debug: true
    hsv_ranges:
      - 15.0
      - 20.0
      - 50.0
      - 200.0
      - 20.0
      - 200.0
```

This node is designed to obtain a result for each image that arrives, so the processing is done directly in the callback, as long as there is a subscriber to this result.

Creating an image_transport::Subscriber is very similar to a rclcpp::Subscription. The first parameter is a rclcpp::Node*, so we use this. The fourth parameter indicates the transport method, in this case raw. We adjust the quality of service in the last parameters to the usual in sensors.

```
src/br2_tracking/ObjectDetector.cpp
```

```cpp
ObjectDetector::ObjectDetector()
: Node("object_detector")
{
  image_sub_ = image_transport::create_subscription(
    this, "input_image", std::bind(&ObjectDetector::image_callback, this, _1),
    "raw", rclcpp::SensorDataQoS().get_rmw_qos_profile());

  detection_pub_ = create_publisher<vision_msgs::msg::Detection2D>("detection", 100);
}

void
ObjectDetector::image_callback(const sensor_msgs::msg::Image::ConstSharedPtr & msg)
{
  if (detection_pub_->get_subscription_count() == 0) {return;}
  ...

  vision_msgs::msg::Detection2D detection_msg;
  ...
  detection_pub_->publish(detection_msg);
}
```

Image processing was already introduced in the previous sections. Once the image message has been transformed to a `cv::Mat`, we proceed to transform it from RGB to HSV, and we do a color filter. The `cv::boundingRect` function calculates a bounding box from the mask resulting from the color filtering. The `cv::moments` function calculates the center of mass of these pixels.

```
src/br2_tracking/ObjectDetector.cpp
```

```cpp
const float & h = hsv_filter_ranges_[0];
const float & H = hsv_filter_ranges_[1];
const float & s = hsv_filter_ranges_[2];
const float & S = hsv_filter_ranges_[3];
const float & v = hsv_filter_ranges_[4];
const float & V = hsv_filter_ranges_[5];

cv_bridge::CvImagePtr cv_ptr;
try {
  cv_ptr = cv_bridge::toCvCopy(msg, sensor_msgs::image_encodings::BGR8);
} catch (cv_bridge::Exception & e) {
  RCLCPP_ERROR(get_logger(), "cv_bridge exception: %s", e.what());
  return;
}

cv::Mat img_hsv;
cv::cvtColor(cv_ptr->image, img_hsv, cv::COLOR_BGR2HSV);

cv::Mat1b filtered;
cv::inRange(img_hsv, cv::Scalar(h, s, v), cv::Scalar(H, S, V), filtered);

auto moment = cv::moments(filtered, true);
cv::Rect bbx = cv::boundingRect(filtered);

auto m = cv::moments(filtered, true);

if (m.m00 < 0.000001) {return;}

int cx = m.m10 / m.m00;
int cy = m.m01 / m.m00;

vision_msgs::msg::Detection2D detection_msg;
detection_msg.header = msg->header;
detection_msg.bbox.size_x = bbx.width;
detection_msg.bbox.size_y = bbx.height;
detection_msg.bbox.center.x = cx;
detection_msg.bbox.center.y = cy;
detection_msg.source_img = *cv_ptr->toImageMsg();
detection_pub_->publish(detection_msg);
```

In the previous code, the image is processed, the bounding box `bbx` of the filtered pixels in `filtered` is obtained, and it is published, together with the center of mass (cx, cy). In addition, the optional `source_img` field is filled in, since we require the size of the image in `object_tracker_main.cpp`.

The `HeadController` implementation is a bit more complex. Let's focus first on the fact that it is a Lifecycle node, and that its control loop is only called when it is active. Let's look at the declaration of the node, just the part of its control infrastructure:

```
include/br2_tracking/HeadController.hpp

class HeadController : public rclcpp_lifecycle::LifecycleNode
{
public:
  HeadController();

  CallbackReturn on_configure(const rclcpp_lifecycle::State & previous_state);
  CallbackReturn on_activate(const rclcpp_lifecycle::State & previous_state);
  CallbackReturn on_deactivate(const rclcpp_lifecycle::State & previous_state);

  void control_sycle();

private:
  rclcpp_lifecycle::LifecyclePublisher<trajectory_msgs::msg::JointTrajectory>::SharedPtr
    joint_pub_;
  rclcpp::TimerBase::SharedPtr timer_;
};
```

The `LifecycleNode::create_subscription` method returns an `rclcpp_lifecycle:: LifecyclePublisher` instead of an `rclcpp::Publisher`. Although its functionality is similar, it is necessary to activate it so that it can be used.

A LifeCycleNode can redefine the functions that are called when a transition between states is triggered in the derived class. These functions can return SUCCESS or FAILURE. If it returns SUCCESS, the transition is allowed. If FAILURE is returned, it is not transitioned to the new state. All of these methods return SUCCESS in the base class, but the developer can redefine them to establish the rejection conditions.

In this case, the transitions leading to the inactive state (`on_configure`) and those that transition between active and inactive (`on_activate` and `on_deactivate`) are redefined:

```
src/br2_tracking/HeadController.cpp
```

```cpp
HeadController::HeadController()
: LifecycleNode("head_tracker")
{
  joint_pub_ = create_publisher<trajectory_msgs::msg::JointTrajectory>(
    "joint_command", 100);
}

CallbackReturn
HeadController::on_configure(const rclcpp_lifecycle::State & previous_state)
{
  return CallbackReturn::SUCCESS;
}

CallbackReturn
HeadController::on_activate(const rclcpp_lifecycle::State & previous_state)
{
  joint_pub_->on_activate();
  timer_ = create_wall_timer(100ms, std::bind(&HeadController::control_sycle, this));
```

```
src/br2_tracking/HeadController.cpp
```

```cpp
  return CallbackReturn::SUCCESS;
}

CallbackReturn
HeadController::on_deactivate(const rclcpp_lifecycle::State & previous_state)
{
  joint_pub_->on_deactivate();
  timer_ = nullptr;

  return CallbackReturn::SUCCESS;
}

void
HeadController::control_sycle()
{
}
```

All previous transitions return SUCCESS, so all transitions are carried out. In the case of developing a laser driver, for example, some transition (configure or activate) would fail if the device is not found or cannot be accessed.

The above code has two aspects that are interesting to explain:

- The `control_cycle` method contains our control logic and is set to run at 10 Hz. Note that the timer is created at `on_activate`, which is when the active state is transitioned. Likewise, disabling this timer is simply destroying it by going inactive. This way `control_cycle` will not be called and the control logic will only be executed when the node is active.

- The publisher must be activated in `on_activate` and deactivated in `on_deactivate`.

The `HeadController` node will execute iteratively, receiving the current state of the neck joints through the topic `/joint_state`, and of the move commands through the `/command` topic. As usual in this schematic, both values in `last_state_` and `last_command_` are stored to be used when we execute the next cycle of the control logic. Also, the timestamp of the last received command is saved. When stopping receiving commands, the robot should return to the initial position.

```
include/br2_tracking/HeadController.hpp

class HeadController : public rclcpp_lifecycle::LifecycleNode
{
public:
  void joint_state_callback(
    control_msgs::msg::JointTrajectoryControllerState::UniquePtr msg);
  void command_callback(br2_tracking_msgs::msg::PanTiltCommand::UniquePtr msg);

private:
  rclcpp::Subscription<br2_tracking_msgs::msg::PanTiltCommand>::SharedPtr command_sub_;
  rclcpp::Subscription<control_msgs::msg::JointTrajectoryControllerState>::SharedPtr
    joint_sub_;
  rclcpp_lifecycle::LifecyclePublisher<trajectory_msgs::msg::JointTrajectory>::SharedPtr
    joint_pub_;

  control_msgs::msg::JointTrajectoryControllerState::UniquePtr last_state_;
  br2_tracking_msgs::msg::PanTiltCommand::UniquePtr last_command_;
  rclcpp::Time last_command_ts_;
};
```

```
src/br2_tracking/HeadController.cpp

void
HeadController::joint_state_callback(
  control_msgs::msg::JointTrajectoryControllerState::UniquePtr msg)
{
  last_state_ = std::move(msg);
}

void
HeadController::command_callback(br2_tracking_msgs::msg::PanTiltCommand::UniquePtr msg)
{
  last_command_ = std::move(msg);
  last_command_ts_ = now();
}
```

The format of `control_msgs::msg::JointTrajectoryControllerState` is designed to report the name of the controlled joints, as well as the desired, current, and error trajectories:

```
$ ros2 interface show control_msgs/msg/JointTrajectoryControllerState

std_msgs/Header header
string[] joint_names
trajectory_msgs/JointTrajectoryPoint desired
trajectory_msgs/JointTrajectoryPoint actual
trajectory_msgs/JointTrajectoryPoint error # Redundant, but useful
```

Using a `trajectory_msgs::msg::JointTrajectory` may seem complicated at first, but it is not if we analyze the following code, which is a command to put the robot's neck in the initial state while looking at Figure 5.6:

```
src/br2_tracking/HeadController.cpp

CallbackReturn
HeadController::on_deactivate(const rclcpp_lifecycle::State & previous_state)
{
  trajectory_msgs::msg::JointTrajectory command_msg;
  command_msg.header.stamp = now();
  command_msg.joint_names = last_state_->joint_names;
  command_msg.points.resize(1);
  command_msg.points[0].positions.resize(2);
  command_msg.points[0].velocities.resize(2);
  command_msg.points[0].accelerations.resize(2);
  command_msg.points[0].positions[0] = 0.0;
  command_msg.points[0].positions[1] = 0.0;
  command_msg.points[0].velocities[0] = 0.1;
  command_msg.points[0].velocities[1] = 0.1;
  command_msg.points[0].accelerations[0] = 0.1;
  command_msg.points[0].accelerations[1] = 0.1;
  command_msg.points[0].time_from_start = rclcpp::Duration(1s);

  joint_pub_->publish(command_msg);

  return CallbackReturn::SUCCESS;
}
```

- The joint_names field is a std::vector<std::string> containing the name of the joints being controlled. In this case, there are two, and they are the same ones that are already in the state message.

- A single waypoint will be sent (for this reason, the points field is resized to 1), in which a position, speed, and acceleration must be specified for each joint (since there are two joints, each of these fields is resized to two). Position 0 corresponds to the joint that in joint_names is at 0, and so on.

- time_from_start indicates the time required to reach the commanded position. As it is the last command sent before deactivating (that is why its desired positions are 0), one second will be enough not to force the neck motors.

The controller of the neck joints is controlled by sending commands containing positions, but what is received from the ObjectDetector is the speed control that should be done to center the detected object in the image.

The first implementation could be to send as position, the current position combined with the received control:

```
src/br2_tracking/HeadController.cpp

command_msg.points[0].positions[0] = last_state_->actual.positions[0] - last_command_->pan;
command_msg.points[0].positions[1] = last_state_->actual.positions[1] -last_command_->tilt;
```

If the reader uses this implementation, he would see that if we want to be reactive enough, even if the difference between the ObjectDetector and HeadDetector frequencies were small, the robot's head might start to oscillate, trying to center the image on the detected object. It is difficult for the robot to maintain a stable focus on the detected object. This problem is solved in engineering using a PID controller, one per joint that limits the speed while also absorbing small unwanted oscillations of the neck.

```
include/br2_tracking/HeadController.hpp

class HeadController : public rclcpp_lifecycle::LifecycleNode
{
private:
  PIDController pan_pid_, tilt_pid_;
};
```

For each PID, define a value for the proportional component K_p, the integrating component K_i, and the derivative component K_d. Without going into details, since it is not the objective of this book to describe in-depth the underlying control theory, intuitively, the proportional component brings us closer to the objective. The integrator component compensates for persistent deviations that move us away from the objective. The derivative component tries to damp minor variations when close to the control objective. Figure 5.10 shows a diagram of this PID controller.

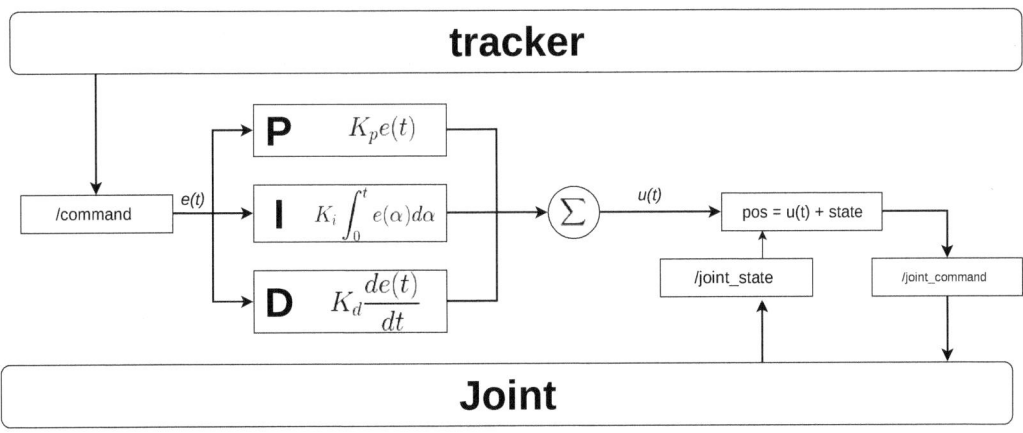

Figure 5.10: Diagram for PID for one joint.

The control command coming from the tracker is the value that the PID should try to keep at 0, so it is the error in t, $e(t)$. Each component of the PID is computed separately and then added to obtain the control to apply $u(t)$. The position sent to the joints will be the current position of the joint plus $u(t)$. The system feeds back since at $t + 1$ the effect of the control is reflected in a change in the object's position in the image toward its center.

Our PID starts by specifying four values: the minimum and maximum input reference expected in the PID and the minimum and maximum output produced. Negative input produces negative outputs:

```
src/br2_tracking/PIDController.hpp

class PIDController
{
public:
  PIDController(double min_ref, double max_ref, double min_output, double max_output);

  void set_pid(double n_KP, double n_KI, double n_KD);
  double get_output(double new_reference);
};
```

```
src/br2_tracking/HeadController.cpp
```

```cpp
HeadController::HeadController()
: LifecycleNode("head_tracker"),
  pan_pid_(0.0, 1.0, 0.0, 0.3),
  tilt_pid_(0.0, 1.0, 0.0, 0.3)
{
}
CallbackReturn
HeadController::on_configure(const rclcpp_lifecycle::State & previous_state)
{
  pan_pid_.set_pid(0.4, 0.05, 0.55);
  tilt_pid_.set_pid(0.4, 0.05, 0.55);
}
void
HeadController::control_sycle()
{
  double control_pan = pan_pid_.get_output(last_command_->pan);
  double control_tilt = tilt_pid_.get_output(last_command_->tilt);

  command_msg.points[0].positions[0] = last_state_->actual.positions[0] - control_pan;
  command_msg.points[0].positions[1] = last_state_->actual.positions[1] - control_tilt;
}
```

5.2.6 Executing the Tracker

In the main program `object_tracker_main.cpp` all the nodes are created and added to an executor. Just before starting spinning the nodes, we trigger the configure transition for node `node_head_controller`. The node will be ready to be activated when requested.

```
src/object_tracker_main.cpp
```

```cpp
rclcpp::executors::SingleThreadedExecutor executor;
executor.add_node(node_detector);
executor.add_node(node_head_controller->get_node_base_interface());
executor.add_node(node_tracker);

node_head_controller->trigger_transition(
  lifecycle_msgs::msg::Transition::TRANSITION_CONFIGURE);
```

A launcher remaps the topics and loads the file with the HSV filter parameters:

```
launch/tracking.launch.py
```

```python
params_file = os.path.join(
  get_package_share_directory('br2_tracking'),
  'config',
  'detector.yaml'
  )

object_tracker_cmd = Node(
  package='br2_tracking',
  executable='object_tracker',
  parameters=[{
    'use_sim_time': True
  }, params_file],
  remappings=[
    ('input_image', '/head_front_camera/rgb/image_raw'),
    ('joint_state', '/head_controller/state'),
    ('joint_command', '/head_controller/joint_trajectory')
  ],
  output='screen'
)
```

Start the Tiago simulated (the home world, by default) gazebo:

```
$ ros2 launch br2_tiago sim.launch.py
```

In another terminal, launch the project:

```
$ ros2 launch br2_tracking tracking.launch.py
```

The detection windows are not shown until the first object is detected, but HeadController is in *Inactive* state, and no tracking will be done.

See how we can manage the LifeCycleNode at runtime, such as head_tracker (the name of the HeadController node). Keep our project running, with the robot tracking an object.

Using the following command, check what LifeCycle nodes are currently running:

```
$ ros2 lifecycle nodes
/head_tracker
```

Now verify the state it is currently in:

```
$ ros2 lifecycle get /head_tracker
inactive [3]
```

Good. The LifeCycleNode is in the *Inactive* state, just as expected. Obtain what transitions can be triggered from the current state:

```
$ ros2 lifecycle list /head_tracker
- cleanup [2]
            Start: inactive
            Goal: cleaningup
- activate [3]
            Start: inactive
            Goal: activating
- shutdown [6]
            Start: inactive
            Goal: shuttingdown
```

Activate the node to start tracking the detected object:

```
$ ros2 lifecycle set /head_tracker activate
Transitioning successful
```

Run a teleoperator in a third terminal to teleoperate the robot toward the furniture. Then, the robot will move (when HeadController is *Active*) the head to center the furniture in the image. As soon as the robot does not perceive the objects, it will move the head to the initial position:

```
$ ros2 run teleop_twist_keyboard teleop_twist_keyboard --ros-args -r
cmd_vel:=key_vel
```

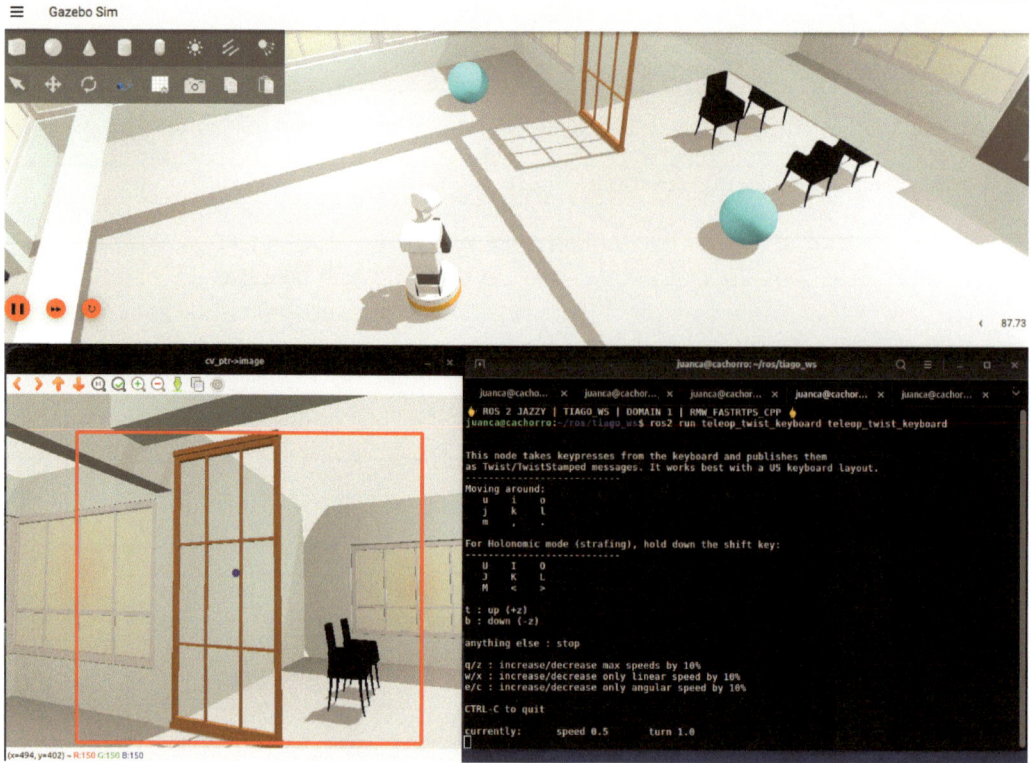

Figure 5.11: Project tracking running.

Deactivate the node and check how the neck of the robot returns to its initial position. Remember that it was what was commanded in this transition, in the on_deactivate method.

```
$ ros2 lifecycle set /head_tracker deactivate

Transitioning successful
```

To activate it again, type:

```
$ ros2 lifecycle set /head_tracker activate

Transitioning successful
```

PROPOSED EXERCISES:

1. In AvoidanceNodeNode, instead of using the nearest obstacle, uses all nearby detected obstacles to compute the repulsion vector.

2. In ObjectDetector, instead of calculating a building block that encloses all the pixels that pass the filter, calculate a bounding box for each independent object. Publish the bounding boxes corresponding to the object most recently detected.

3. Try to make HeadController more reactive.

## CHAPTER 6

Programming Robot Behaviors with Behavior Trees

BEHAVIOR Trees for robot control [8] have become very popular in recent years. They have been used in various applications, mainly in video games and robots. They are usually compared to finite state machines, but the reality is that they are different approximations. When developing robotic behaviors with FSMs (Finite State Machines), we think about states and transitions. When we use Behavior Trees, we think of sequences, fallbacks, and many flow resources that give them great expressiveness. In this chapter, as an illustrative example, we will implement the *Bump and Go* that we did with FSMs in the Chapter 3, and we will see how much the two approaches differ.

6.1 BEHAVIOR TREES

A Behavior Tree (BT) is a mathematical model to encode the control of a system. A BT is a way to structure the switching between different tasks in an autonomous agent, such as a robot or a virtual entity in a computer game. It is a hierarchical data structure defined recursively from a root node with several child nodes. Each child node, in turn, can have more children, and so on. Nodes that do not have children are usually called leaves of the tree.

The basic operation of a node is the **tick**. When a node is ticked, it can return three different values:

- **SUCCESS**: The node has completed its mission successfully.

- **FAILURE**: The node has failed in its mission.

- **RUNNING**: The node has not yet completed its mission.

A BT has four different types of nodes:

DOI: 10.1201/9781003516798-6

- **Control**: These types of nodes have 1-N children. Its function is to spread the tick to their children.

- **Decorators**: They are control nodes with only one child.

- **Action**: They are the leaves of the tree. The user must implement action nodes since they must generate the control required by the application.

- **Condition**: They are action nodes that cannot return RUNNING. In this case, the value. SUCCESS is understood as the condition it encodes is met, and FAILURE if it is not.

Figure 6.1 shows a simple BT. When a BT is executed, the root node is ticked until it finishes executing, that is, until it returns SUCCESS or FAILURE.

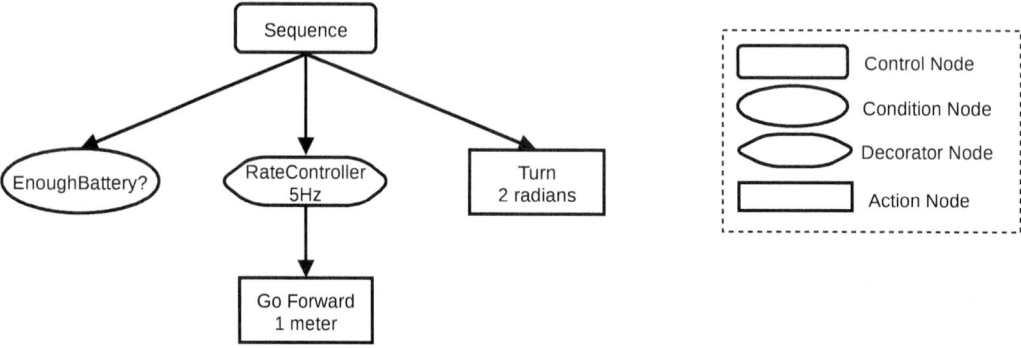

Figure 6.1: Simple Behavior Tree with various types of Nodes.

- The root node is a control node of type `Sequence`. This node ticks its children in order, starting from left. When a child returns SUCCESS, the sequence node ticks the next one. If the child node returns something else, the sequence node returns this value.

- The first child, `EnoughtBattery?`, is a Condition node. If it returns SUCCESS, it indicates that there is enough battery for the robot to carry out its mission so that the sequence node can advance to the next child. If it returned FAILURE, the mission would be aborted, as the result of executing the BT would be FAILURE.

- The `Go Forward` action node commands the robot to advance. As long as it has not traveled 1 m, the node returns RUNNING with each tick. When it has traveled the specified distance, it will return SUCCESS.

- The `Go Forward` action node has as its parent a Decorator node that controls that the frequency at which its child ticks is not greater than 5 Hz. Meanwhile, each tick returns the value returned by the child in the last tick.

- The `Turn` action node is similar to `Go Forward`, but spinning the robot 2 radians.

The library of available nodes can be extended with nodes created by the user. As we have said before, the user must implement the action nodes, but if we need any other type of node that is not available, we can implement it. In the above example, the `RateController` decorator node is not part of the Behavior Tree core library but can be implemented by the user.

A Behavior Tree controls the action decision flow. Leaves are not intended to implement complex algorithms or subsystems. The BT leaves should coordinate other subsystems in the robot. In ROS 2, this is done by publishing or subscribing to topics or using ROS 2 services/actions. Figure 6.2 shows a BT in which the nodes are used to coordinate the actions of a robot. Observe how the complexity is in the subsystem that coordinates, not in the BT leaves.

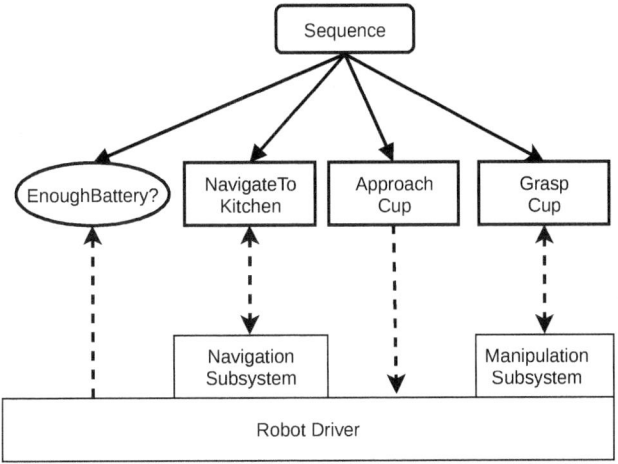

Figure 6.2: BT where the leaves control a robot by publish/subscribe (one-way dotted arrow) or ROS 2 actions (two-way dotted arrow).

The second control node that we will present is the `Fallback`. This node can express fallback strategies, that is, what to do if a node returns FAILURE. Figure 6.3 shows an example of the use of this node.

1. The `Fallback` node ticks the first child. If it returns FAILURE, it ticks the next child.

2. If the second child returns SUCCESS, the `Fallback` node returns SUCCESS. Otherwise, it ticks the next child.

3. If all children have returned FAILURE, the `Fallback` node returns FAILURE.

In the development cycle with Behavior Trees, we can identify two phases:

- **Node Development**: Action nodes and any other node that the user requires for their application are designed, developed, and compiled in this phase. These nodes become part of the library of available nodes at the same category as the core nodes of Behavior Trees.

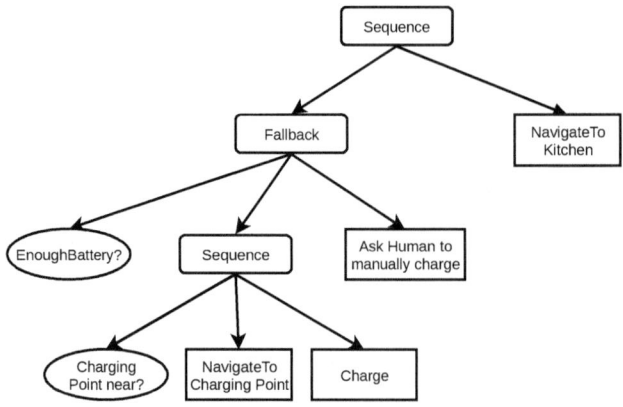

Figure 6.3: BT with a fallback strategy for charging battery.

- **Deployment**: In this phase, the Behavior Tree is composed using the available nodes. It is important to note that multiple different Behavior Trees can be created with the same nodes. If the nodes have been designed sufficiently generally, in this phase, very different behaviors of the robot can be defined using the same nodes.

A Behavior Tree has a *blackboard*, a key/value storage that all nodes in a tree can access. Nodes can have input ports and output ports to exchange information between them. The output ports of one node are connected to the input ports of another node using a key from the blackboard. While the ports of the nodes (their type and port class) have to be known at compile-time, the connections are established at deployment-time.

Figure 6.4 shows an example of connecting nodes through ports. A `DetectObject` action node is in charge of detecting some object so that the `InformHuman` node communicates it to the robot operator. `DetectObject` uses its output port `detected_id` to send the identifier of the detected object to `InformHuman` through its port `object_id`. For this, they use the input of the blackboard whose key `objID` currently has the value `cup`. Using keys from the blackboard is not mandatory. At deployment time, the value could be a constant value.

Behavior Trees are specified in XML. Although editing tools such as Groot[1] are used, they generate a BT in XML format. If this BT is saved to disk and this file is loaded from an application, any change to the BT does not require recompiling. The format is easy to understand, and it is common for BTs to be designed directly in XML. The following code shows two equally valid options for the BT in Figure 6.1.

---

[1]`https://github.com/BehaviorTree/Groot`

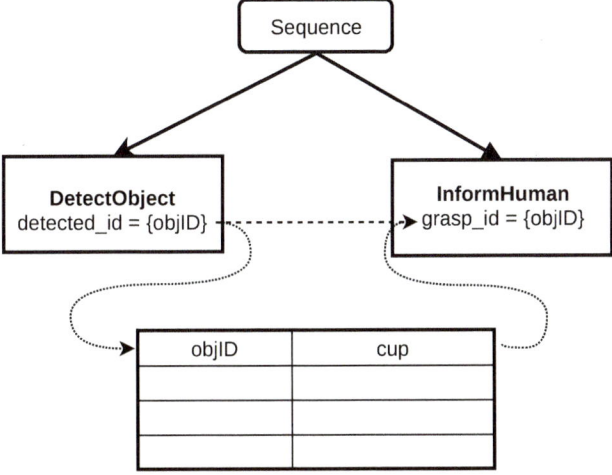

Figure 6.4: Ports connection using a blackboard key.

Compact XML syntax

```
<BehaviorTree ID="BehaviorTree">
    <Sequence>
        <EnoughBattery/>
        <RateController Rate="5Hz">
            <GoForward distance="1.0"/>
        </RateController>
        <Turn angle="2.0"/>
    </Sequence>
</BehaviorTree>
```

Extended XML syntax

```
<?xml version="1.0"?>
<root main_tree_to_execute="BehaviorTree">

    <BehaviorTree ID="BehaviorTree">
        <Sequence>
            <Condition ID="EnoughBattery"/>
            <Decorator ID="RateController" Rate="5Hz">
                <Action ID="GoForward" distance="1.0"/>
            </Decorator>
            <Action ID="Turn" angle="2.o"/>
        </Sequence>
    </BehaviorTree>

    <TreeNodesModel>
        <Condition ID="EnoughBattery"/>
        <Action ID="GoForward">
            <input_port name="distance"/>
        </Action>
        <Decorator ID="RateController">
            <input_port name="Rate"/>
        </Decorator>
        <Action ID="Turn">
            <input_port name="angle"/>
        </Action>
    </TreeNodesModel>
</root>
```

Table 6.1 shows a summary of the commonly available control nodes. This table shows what a control node returns when ticked, based on what the ticked child

Control Node Type	Value returned by child		
	FAILURE	**SUCCESS**	**RUNNING**
Sequence	Return FAILURE and restart sequence	Tick next child. Return SUCCESS if no more child	Return RUNNING and tick again
ReactiveSequence	Return FAILURE and restart sequence	Tick next child. Return SUCCESS if no more child	Return RUNNING and restart sequence
SequenceStar	Return FAILURE and tick again	Tick next child. Return SUCCESS if no more child	Return RUNNING and tick again
Fallback	Tick next child. Return FAILURE if no more child	Return SUCCESS	Return RUNNING and tick again
ReactiveFallback	Tick next child. Return FAILURE if no more child	Return SUCCESS	Return RUNNING and restart sequence
InverterNode	Return SUCCESS	Return FAILURE	Return RUNNING
ForceSuccessNode	Return SUCCESS	Return SUCCESS	Return RUNNING
ForceFailureNode	Return FAILURE	Return FAILURE	Return RUNNING
RepeatNode (N)	Return FAILURE	Return RUNNING N times before returning SUCCESS	Return RUNNING
RetryNode (N)	Return RUNNING N times before returning FAILURE	Return SUCCESS	Return RUNNING

Table 6.1: Summary of the behavior of the control nodes. Cell color groups into sequence, fallback, and decorator nodes.

returns. In the case of sequences and fallbacks, it also shows what it does if this control node is ticked again: tick the next, restart the first child, or insist on the same child.

Let's analyze in detail some of these control nodes:

- **Sequence nodes**: In the previous section, we have used the basic sequence node. Behavior Trees allow sequence nodes with different behavior, which is helpful in some applications.

  - **Sequence**: As explained in the previous section, this node ticks its first child. When it returns SUCCESS, the ticks are made to the next child, and so on. If any child returns FAILURE, this node returns FAILURE and, if ticked again, starts over from the first child.

    Figure 6.5 shows an example of a sequence in which to take an image, it must check that the object is close and that the camera is ready. Once the camera is pointed at the subject, a picture can be taken. If any of the above children fail, the sequence fails. No child repeats its execution if it has already indicated that it has finished successfully.

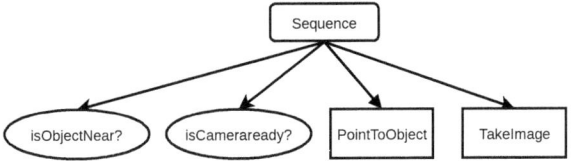

Figure 6.5: Example of Sequence node.

– **ReactiveSequence**: this sequence is commonly used when it is necessary to check conditions continuously. If any child returns RUNNING, the sequence restarts from the beginning. In this way, all nodes are always ticked up to the one returned by RUNNING on the previous tick.

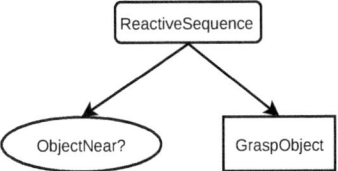

Figure 6.6: Example of ReactiveSequence node.

– **SequenceStar**: This sequence is used to avoid restarting a sequence if, at some point, it has returned a FAILURE child. If this sequence is ticked again after a failure, the failed node is ticked directly.

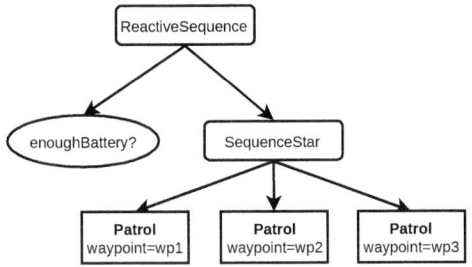

Figure 6.7: Example of ReactiveStar node.

• **Fallback nodes**: As we presented previously, fallback nodes allow us to execute different strategies to satisfy a condition until we find a successful one.

– **Fallback**: It is the basic version of this control node. The children tick in sequence. When one returns FAILURE, it moves on to the next. The moment one returns SUCCESS, the fallback node returns SUCCESS.

– **ReactiveFallback**: This alternative version of fallback has the difference that if a node returns RUNNING, the sequence is restarted from the beginning. The next tick will be made again to the first child. It is useful when the first node is a condition, and it must be checked while executing the action that tries to satisfy it. For example, in Figure 6.8, the action of charging the robot is running while the battery is not charged.

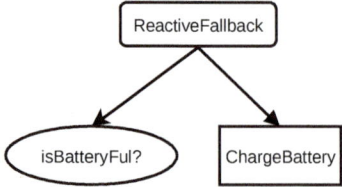

Figure 6.8: Example of ReactiveFallback node.

- **Decorator nodes**: They modify the return value of their only child. In the case of `RepeatNode` and `RetryNode`, they receive the `N` repetitions or retries through their input port.

6.2 *BUMP AND GO* WITH BEHAVIOR TREES

In this section we will show how to implement action nodes within our ROS 2 packages, and how these nodes can access the Computation Graph to communicate with other nodes. To do this, we will reimplement the *Bump and Go* example that we did with state machines in Chapter 3, and thus we will see the differences that exist.

Let's start with the design of the Behavior Tree (Figure 6.10). It seems clear that we will need the following BT nodes (Figure 6.9):

- A condition node that indicates whether there is an obstacle (SUCCESS) or not (FAILURE) depending on the information received from the laser.

- Three action nodes that make the robot turn, move or go forward publishing speed messages. `Back` and `Turn` will return RUNNING for 3 s before returning SUCCESS. `Forward` will return RUNNING in all ticks.

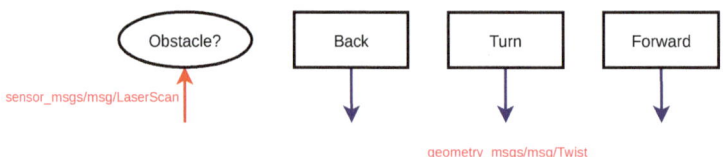

Figure 6.9: Action nodes for *Bump and Go*.

The Computation Graph is similar to the one in Figure 3.4, so we will skip its explanation. Let's focus on the workspace:

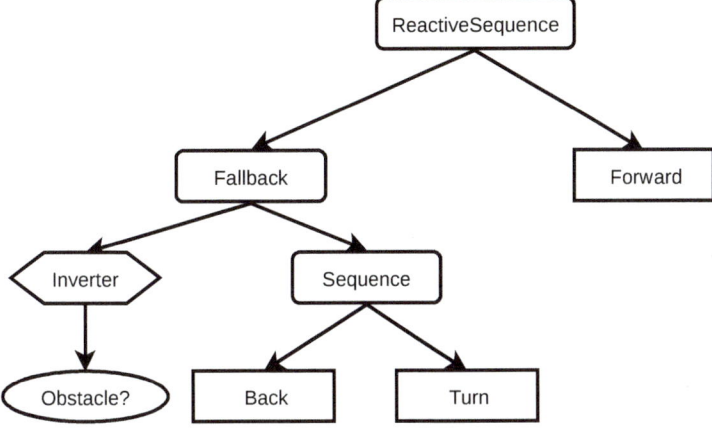

Figure 6.10: Complete Behavior Tree for *Bump and Go*.

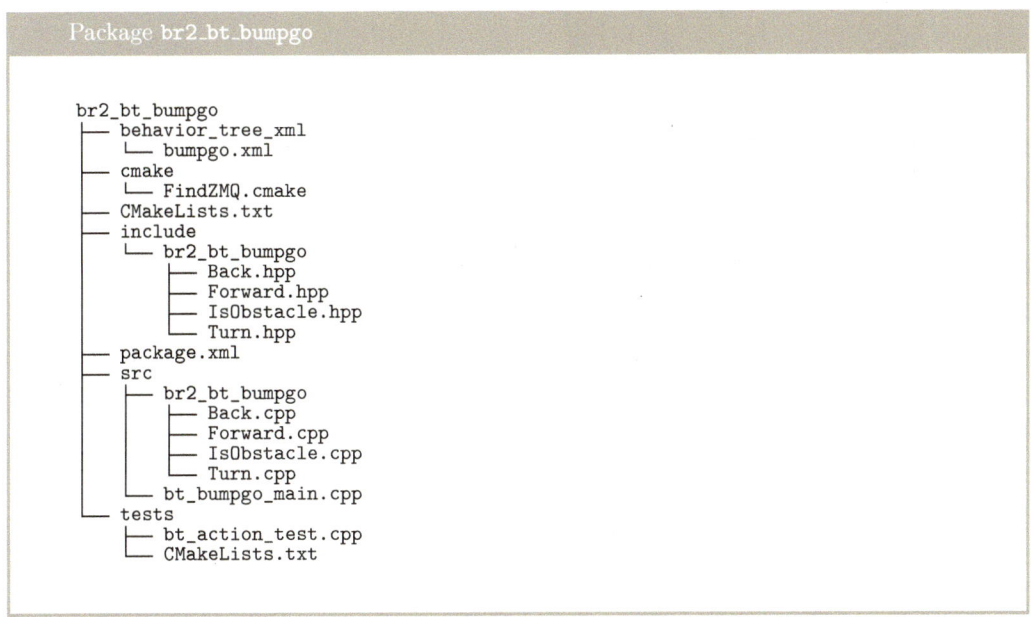

- Each of the BT nodes is a C++ class. Just like when we implement ROS 2 nodes, we create a directory with the package name in `src` for sources and a directory with the same name in `include` for headers.

- A `tests` directory where there will be tests with gtest and a program to manually test a BT node, as we will explain later.

- A `cmake` directory contains a cmake file to find the ZMQ[2] library needed to debug Behavior Trees at runtime.

- A `behavior_tree_xml` directory with XML files containing the structure of the behavior trees that we will use in this package.

---

[2]`https://zeromq.org`

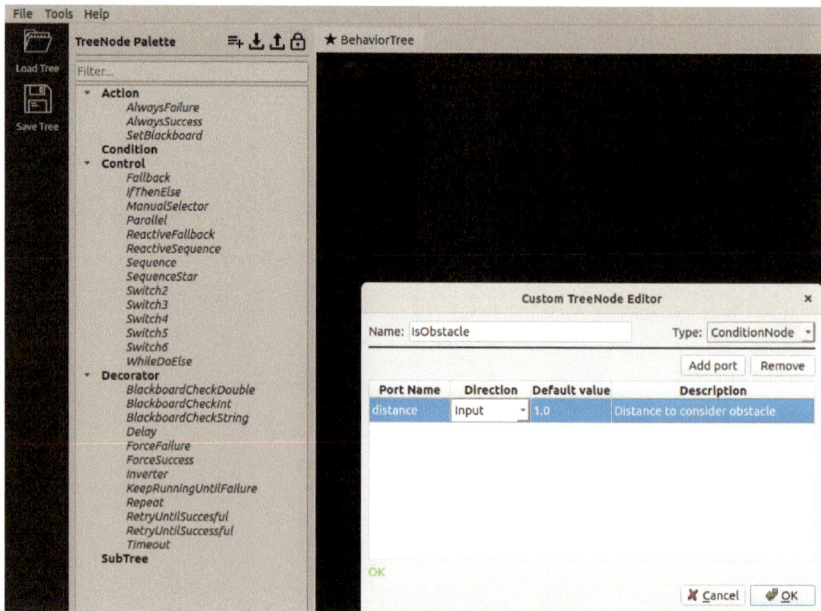

Figure 6.11: Specification of IsObstacle BT node.

6.2.1 Using Groot to Create the Behavior Tree

This section introduces a tool for developing and monitoring Behavior Trees, which is Groot. The behavior trees in this package are already created, but we find it helpful to explain how this tool works. It is useful for monitoring runtime performance, or perhaps the reader wants to make modifications.

Groot is included in the repository dependencies, so to execute it, simply type:

```
$ ros2 run groot Groot
```

After selecting the editor, follow next steps:

- Add the nodes `Turn, Forward, Back`, and `IsObstacle` to the palette. All are Action Nodes except `IsObstacle`, for which add an input port, as shown in Figure 6.11.

- Save the palette.

- Create the Behavior Tree as shown in Figure 6.12.

- Save the Behavior Tree in `mr2_bt_bumpgo/behavior_tree_xml/bumpgo.xml`.

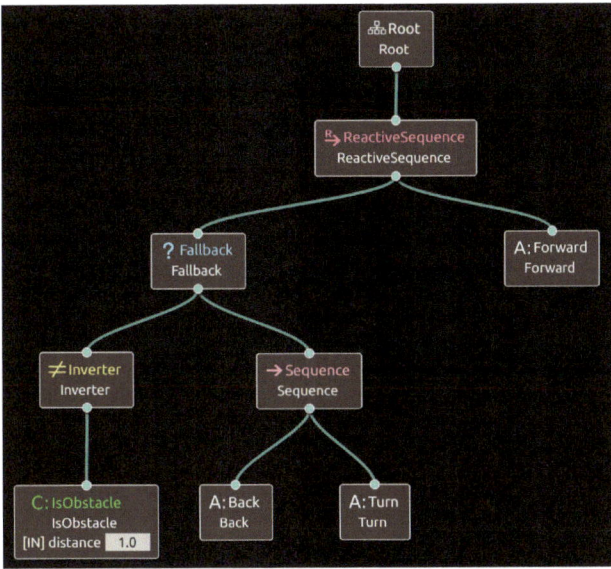

Figure 6.12: Action nodes for *Bump and Go*.

```
behavior_tree_xml/bumpgo.xml
```

```xml
<?xml version="1.0"?>
<root main_tree_to_execute="BehaviorTree">
    <BehaviorTree ID="BehaviorTree">
        <ReactiveSequence>
            <Fallback>
                <Inverter>
                    <Condition ID="IsObstacle" distance="1.0"/>
                </Inverter>
                <Sequence>
                    <Action ID="Back"/>
                    <Action ID="Turn"/>
                </Sequence>
            </Fallback>
            <Action ID="Forward"/>
        </ReactiveSequence>
    </BehaviorTree>
    <TreeNodesModel>
        <Action ID="Back"/>
        <Action ID="Forward"/>
        <Condition ID="IsObstacle">
            <input_port default="1.0" name="distance">Dist to consider obst</input_port>
        </Condition>
        <Action ID="Turn"/>
    </TreeNodesModel>
</root>
```

The Behavior Tree specification in XML is straightforward. There are two parts:

- **BehaviorTree**: It is the specification of the tree structure. The XML tags match the type of BT node specified, and the child nodes are within their parents.

- **TreeNodesModel**: Define the custom nodes we have created, indicating their input and output ports.

There is a valid alternative to this structure, which is ignoring the **TreeNodes-Model** and directly using the name of the custom BT nodes:

```xml
<?xml version="1.0"?>
<root main_tree_to_execute="BehaviorTree">
    <BehaviorTree ID="BehaviorTree">
        <ReactiveSequence>
            <Fallback>
                <Inverter>
                    <IsObstacle distance="1.0"/>
                </Inverter>
                <Sequence>
                    <Back/>
                    <Turn/>
                </Sequence>
            </Fallback>
            <Forward/>
        </ReactiveSequence>
    </BehaviorTree>
</root>
```

6.2.2 BT Nodes Implementation

We will use the Behavior Trees library behaviortree.CPP[3], which is pretty standard in ROS/ROS 2. Let's look at the `Forward` implementation to get an idea of how simple it is to implement a BT node:

```cpp
include/mr2_bt_bumpgo/Forward.hpp

class Forward : public BT::ActionNodeBase
{
public:
  explicit Forward(
    const std::string & xml_tag_name,
    const BT::NodeConfiguration & conf);

  BT::NodeStatus tick();

  static BT::PortsList providedPorts()
  {
    return BT::PortsList({});
  }

private:
  rclcpp::Node::SharedPtr node_;
  rclcpp::Time start_time_;
  rclcpp::Publisher<geometry_msgs::msg::Twist>::SharedPtr vel_pub_;
};
```

As shown in the previous code, when a Behavior Tree is created, an instance of each class of a BT node is constructed for each one that appears in the Behavior Tree. An action node inherits from `BT::ActionNodeBase`, having to implement three methods and setting the constructor arguments:

- The constructor receives the content of the name field (which is optional) in the XML, as well as a `BT::NodeConfiguration` that contains, among other things, a pointer to the blackboard shared by all the nodes of a tree.

- The `halt` method is called when the tree finishes its execution, and it is used

---

[3]https://www.behaviortree.dev

to carry out any cleanup that the node requires. We will define void, as it is a pure virtual method.

- The `tick` method implements the tick operation that we have already described in this chapter.

- A static method that returns the ports of the node. In this case, `Forward` has no ports, so we return an empty list of ports.

The class definition is also straightforward:

```
src/mr2_bt_bumpgo/Forward.cpp

Forward::Forward(
  const std::string & xml_tag_name,
  const BT::NodeConfiguration & conf)
: BT::ActionNodeBase(xml_tag_name, conf)
{
  config().blackboard->get("node", node_);

  vel_pub_ = node_->create_publisher<geometry_msgs::msg::Twist>("/output_vel", 100);
}

BT::NodeStatus
Forward::tick()
{
  geometry_msgs::msg::Twist vel_msgs;

  vel_msgs.linear.x = 0.3;
  vel_pub_->publish(vel_msgs);

  return BT::NodeStatus::RUNNING;
}

}  // namespace br2_bt_bumpgo


#include "behaviortree_cpp_v3/bt_factory.h"
BT_REGISTER_NODES(factory)
{
  factory.registerNodeType<br2_bt_bumpgo::Forward>("Forward");
}
```

- In the constructor, after calling the constructor of the base class, we will get the pointer to the ROS 2 node of the blackboard. We will see soon that when the tree is created, the pointer to the ROS 2 node is inserted into the blackboard with the key "node" so that it is available to any BT node that requires it to create publishers, subscribers, get the time, or any related task to ROS 2.

- The `tick` method is quite obvious: each time the node is ticked, it publishes a speed message to go forward, and return RUNNING.

- In the last part of the previous code, we register this class as implementing the `Forward` BT node. This part will be used when creating the tree.

Once the BT node `Forward` has been analyzed, the rest of the nodes are implemented similarly. Let's see some peculiarities:

- The BT node `Turn` performs its task for 3 s, so it saves the timestamp of its first tick, which is identifiable because its state is still IDLE:

```
src/mr2_bt_bumpgo/Turn.cpp

BT::NodeStatus
Turn::tick()
{
  if (status() == BT::NodeStatus::IDLE) {
    start_time_ = node_->now();
  }

  geometry_msgs::msg::Twist vel_msgs;
  vel_msgs.angular.z = 0.5;
  vel_pub_->publish(vel_msgs);

  auto elapsed = node_->now() - start_time_;

  if (elapsed < 3s) {
    return BT::NodeStatus::RUNNING;
  } else {
    return BT::NodeStatus::SUCCESS;
  }
}
```

- The BT node `isObstacle` saves the laser readings and compares them to the distance set on its input port:

```
src/mr2_bt_bumpgo/isObstacle.cpp

void
IsObstacle::laser_callback(sensor_msgs::msg::LaserScan::UniquePtr msg)
{
  last_scan_ = std::move(msg);
}

BT::NodeStatus
IsObstacle::tick()
{
  double distance = 1.0;
  getInput("distance", distance);

  if (last_scan_->ranges[last_scan_->ranges.size() / 2] < distance) {
    return BT::NodeStatus::SUCCESS;
  } else {
    return BT::NodeStatus::FAILURE;
  }
}
```

Each of the BT nodes will be compiled as a separate library. Later we will see that, when creating the Behavior Tree that contains them, we can load these libraries as plugins, quickly locating the implementation of the custom BT nodes.

```
CMakeLists.txt
```

```
add_library(br2_forward_bt_node SHARED src/br2_bt_bumpgo/Forward.cpp)
add_library(br2_back_bt_node SHARED src/br2_bt_bumpgo/Back.cpp)
add_library(br2_turn_bt_node SHARED src/br2_bt_bumpgo/Turn.cpp)
add_library(br2_is_obstacle_bt_node SHARED src/br2_bt_bumpgo/IsObstacle.cpp)

list(APPEND plugin_libs
  br2_forward_bt_node
  br2_back_bt_node
  br2_turn_bt_node
  br2_is_obstacle_bt_node
)

foreach(bt_plugin ${plugin_libs})
  ament_target_dependencies(${bt_plugin} ${dependencies})
  target_compile_definitions(${bt_plugin} PRIVATE BT_PLUGIN_EXPORT)
endforeach()

install(TARGETS
  ${plugin_libs}
  ARCHIVE DESTINATION lib
  LIBRARY DESTINATION lib
  RUNTIME DESTINATION lib/${PROJECT_NAME}
)
```

6.2.3 Running the Behavior Tree

Running a Behavior Tree is easy. A program should build a tree and start ticking its root until it returns SUCCESS. Behavior trees are created using a BehaviorTreeFactory, specifying an XML file or directly a string that contains the XML. BehaviorTreeFactory needs to load the libraries of the custom nodes as plugins and needs the blackboard to be shared among the BT nodes.

To integrate behavior trees with ROS 2, create a ROS 2 node and put it on the blackboard. As shown before, BT nodes can extract it from the blackboard to create publishers/subscribers or clients/servers of services or actions. Along with the tick at the root of the tree, a spin_some manages the arrival of messages to the ROS 2 node.

See how it looks like the program that carries out the tree creation and execution:

```
src/mr2_bt_bumpgo/isObstacle.cpp
```

```
int main(int argc, char * argv[])
{
  rclcpp::init(argc, argv);

  auto node = rclcpp::Node::make_shared("patrolling_node");

  BT::BehaviorTreeFactory factory;
  BT::SharedLibrary loader;

  factory.registerFromPlugin(loader.getOSName("br2_forward_bt_node"));
  factory.registerFromPlugin(loader.getOSName("br2_back_bt_node"));
  factory.registerFromPlugin(loader.getOSName("br2_turn_bt_node"));
  factory.registerFromPlugin(loader.getOSName("br2_is_obstacle_bt_node"));

  std::string pkgpath = ament_index_cpp::get_package_share_directory("br2_bt_bumpgo");
  std::string xml_file = pkgpath + "/behavior_tree_xml/bumpgo.xml";
```

```
src/mr2_bt_bumpgo/isObstacle.cpp

  auto blackboard = BT::Blackboard::create();
  blackboard->set("node", node);
  BT::Tree tree = factory.createTreeFromFile(xml_file, blackboard);

  auto publisher_zmq = std::make_shared<BT::PublisherZMQ>(tree, 10, 1666, 1667);

  rclcpp::Rate rate(10);

  bool finish = false;
  while (!finish && rclcpp::ok()) {
    finish = tree.rootNode()->executeTick() != BT::NodeStatus::RUNNING;

    rclcpp::spin_some(node);
    rate.sleep();
  }

  rclcpp::shutdown();
  return 0;
}
```

1. At the beginning of the main function, we create a generic ROS 2 node which we then insert into the blackboard. This is the node that we have seen that is pulled from the blackboard in `Forward` to create the speed message publisher.

2. The tree is created by a `BT::BehaviorTreeFactory` from an XML, BT action nodes that we will be implemented, and a blackboard.

   (a) As we will see below, each BT node will be compiled as an independent library. The `loader` object helps to find the library in the system to load the BT Node as a plugin. The `BT_REGISTER_NODES` macro that we saw earlier in the BT nodes definition allows the BT node name to be connected with its implementation within the library.

   ```
   BT::BehaviorTreeFactory factory;
   BT::SharedLibrary loader;

   factory.registerFromPlugin(loader.getOSName("br2_forward_bt_node"));
   ```

   (b) Function `get_package_share_directory` from package `ament_index_cpp` lets to obtain the full path of installed package, in order to read any file within. Remember that this is a package included in the package dependencies.

   ```
   std::string pkgpath = ament_index_cpp::get_package_share_directory(
     "br2_bt_bumpgo");
   std::string xml_file = pkgpath + "/behavior_tree_xml/forward.xml";
   ```

   (c) Finally, after creating the blackboard and inserting the shared pointer to the ROS 2 node there, the factory builds the tree to execute.

   ```
   auto blackboard = BT::Blackboard::create();
   blackboard->set("node", node);
   BT::Tree tree = factory.createTreeFromFile(xml_file, blackboard);
   ```

(d) To debug the Behavior Tree at runtime, create `PublisherZMQ` object that publishes all the necessary information. To create it, indicate the tree, the maximum messages per second, and the network ports to use.

```
auto publisher_zmq = std::make_shared<BT::PublisherZMQ>(
    tree, 10, 1666, 1667);
```

3. In this last part, the tree's root is ticked at 10 Hz while the tree returns RUN-NING while handling any pending work in the node, such as the delivery of messages that arrive at subscribers.

Once compiled, execute the simulator and the node and run the program. The robot should move forward.

```
$ ros2 launch br2_tiago sim.launch.py
```

```
$ ros2 run br2_bt_bumpgo bt_bumpgo --ros-args -r input_scan:=/scan_raw -r
output_vel:=/key_vel -p use_sim_time:=true
```

During program execution, it is possible to use Groot to monitor the state of the Behavior Tree to know which nodes are being ticked and the values they return. Simply boot up Groot and select Monitor instead of Editor. Once pressed connect, monitor the execution, as shown in Figure 6.13.

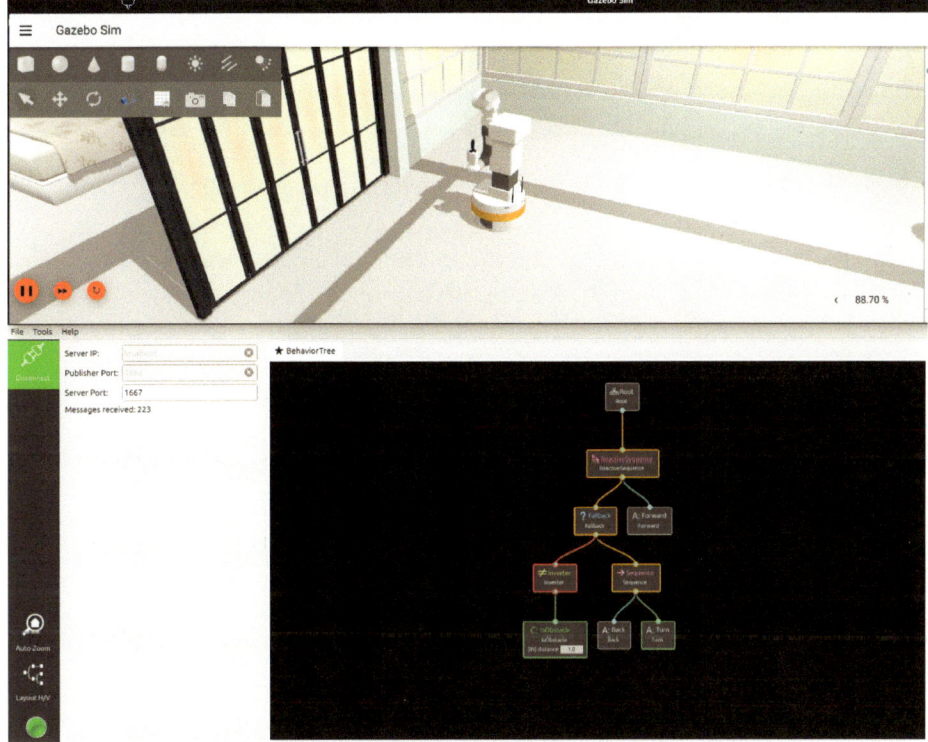

Figure 6.13: Monitoring the execution of a Behavior Tree with Groot.

6.2.4 Testing the BT Nodes

Two types of tests have been included in this package that has been useful during this project's development. They are all in the tests directory of the package.

The first type of test has been to manually test each node separately, running behavior trees that only contain one type of node to see if they work correctly in isolation. We have included only the verification of the BT node Forward:

```
tests/bt_forward_main.cpp

  factory.registerFromPlugin(loader.getOSName("br2_forward_bt_node"));

  std::string xml_bt =
    R"(
    <root main_tree_to_execute = "MainTree" >
      <BehaviorTree ID="MainTree">
          <Forward />
      </BehaviorTree>
    </root>)";

  auto blackboard = BT::Blackboard::create();
  blackboard->set("node", node);
  BT::Tree tree = factory.createTreeFromText(xml_bt, blackboard);

  rclcpp::Rate rate(10);
  bool finish = false;
  while (!finish && rclcpp::ok()) {
    finish = tree.rootNode()->executeTick() != BT::NodeStatus::RUNNING;

    rclcpp::spin_some(node);
    rate.sleep();
  }
```

Start the simulator and run:

```
$ build/br2_bt_bumpgo/tests/bt_forward --ros-args -r input_scan:=/scan_raw -r
output_vel:=/key_vel -p use_sim_time:=true
```

Check that the robot will go forward forever. Do the same with the rest of the BT nodes.

The second type of test is the one recommended in the previous chapter, which is using GoogleTest. It is easy to define a ROS 2 node that records what speeds have been sent to the speed topic.

```
tests/bt_action_test.cpp
```

```cpp
class VelocitySinkNode : public rclcpp::Node
{
public:
  VelocitySinkNode()
  : Node("VelocitySink")
  {
    vel_sub_ = create_subscription<geometry_msgs::msg::Twist>(
      "/output_vel", 100, std::bind(&VelocitySinkNode::vel_callback, this, _1));
  }

  void vel_callback(geometry_msgs::msg::Twist::SharedPtr msg)
  {
    vel_msgs_.push_back(*msg);
  }

  std::list<geometry_msgs::msg::Twist> vel_msgs_;

private:
  rclcpp::Subscription<geometry_msgs::msg::Twist>::SharedPtr vel_sub_;
};
```

It is possible to execute a tree for a few cycles, checking that the speeds that were sent were correct:

```
tests/bt_action_test.cpp
```

```cpp
TEST(bt_action, forward_btn)
{
  auto node = rclcpp::Node::make_shared("forward_btn_node");
  auto node_sink = std::make_shared<VelocitySinkNode>();

  // Creation the Behavior Tree only with the Forward BT node

  rclcpp::Rate rate(10);
  auto current_status = BT::NodeStatus::FAILURE;
  int counter = 0;
  while (counter++ < 30 && rclcpp::ok()) {
    current_status = tree.rootNode()->executeTick();
    rclcpp::spin_some(node_sink);
    rate.sleep();
  }

  ASSERT_EQ(current_status, BT::NodeStatus::RUNNING);
  ASSERT_FALSE(node_sink->vel_msgs_.empty());
  ASSERT_NEAR(node_sink->vel_msgs_.size(), 30, 1);

  geometry_msgs::msg::Twist & one_twist = node_sink->vel_msgs_.front();

  ASSERT_GT(one_twist.linear.x, 0.1);
  ASSERT_NEAR(one_twist.angular.z, 0.0, 0.0000001);
}
```

In this case, after ticking the root of the tree 30 times, see how the node is still returning RUNNING, 30 speed messages have been advertised, and the speeds are correct (they move the robot forward). We could have examined all of them, but we have only done this case for the first one.

Examine the tests of the other nodes. In the case of Turn and Back, it is checked that they do so for the appropriate time before returning success. In the case of isObstacle, we create synthetic laser readings to see if the output is correct in all cases.

6.3   PATROLLING WITH BEHAVIOR TREES

In this section, we will address a more complex and ambitious project. We have previously said that Behavior Tree action nodes help control other subsystems. In the project of the previous section, we have done it in a pretty basic way, processing sensory information and sending speeds. In this section, we will carry out a project in which a Behavior Tree will control more complex subsystems, such as the Nav2 Navigation subsystem and the active vision subsystem that we developed in the previous chapter.

The goal of the project in this section is that of a robot patrolling the simulated house in Gazebo:

- The robot patrols three waypoints in the house (Figure 6.14). Upon reaching each waypoint, the robot turns on itself for a few seconds to perceive its surroundings.

- While the robot goes from one waypoint to another, the robot perceives and tracks the detected objects.

- The robot keeps track (simulated) of the battery level it has. When low, it goes to a recharge point to recharge for a few seconds.

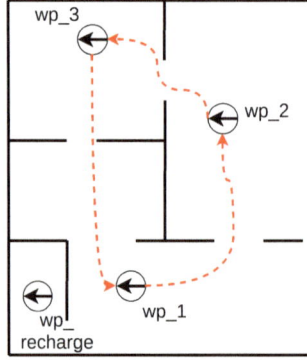

Figure 6.14: Waypoints at the simulated home, with the path followed during patrolling.

Since we are using such a complex and important subsystem as Nav2, the navigation system in ROS 2, we will first describe it in Section 6.3.1. The Section 6.3.2 describes the steps to set up Nav2 for a particular robot and environment. It is possible to skip this section since the br2_navigation package already contains the environment map and configuration files for the simulated Tiago scenario in the house. The following sections already focus on implementing the Behavior Tree and the patrolling nodes.

6.3.1 Nav2 Description

Nav2[4] [9] is the ROS 2 navigation system designed to be modular, configurable, and scalable. Like its predecessor in ROS, it aspires to be the most widely used navigation software, so it supports major robot types: holonomic, differential-drive, legged, and Ackermann (car-like) while allowing information from lasers and 3D cameras to be merged, among others. Nav2 incorporates multiple plugins for local and global navigation and allows custom plugins to be easily used.

The inputs to Nav2 are TF transformations (conforming to REP-105), a map[5], any relevant sensor data sources. It also requires the navigation logic, coded as a BT XML file coded, adapting it to specific problems if needed. Nav2 outputs are the speed sent to the robot base.

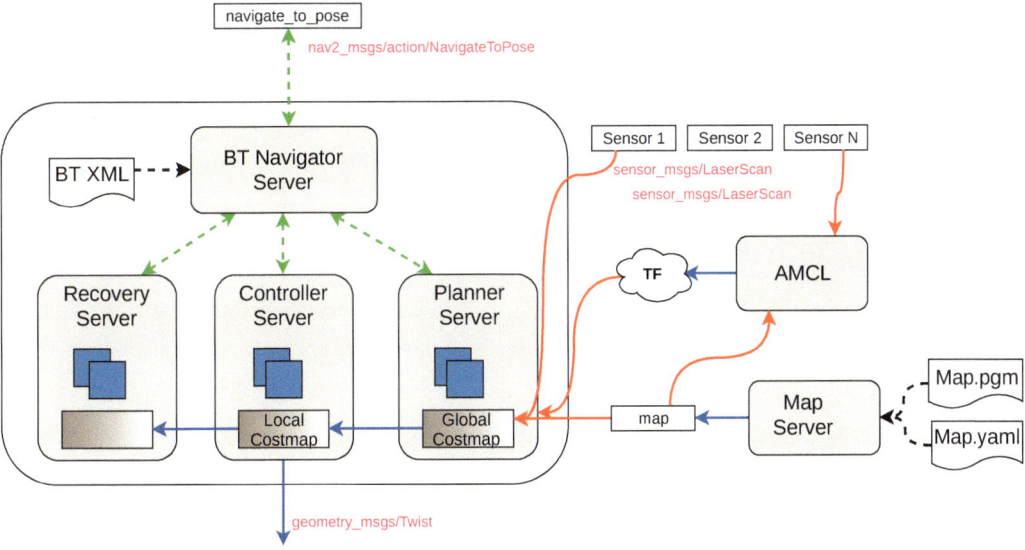

Figure 6.15: Waypoints at the simulated home, with the path followed during patrolling.

Nav2 has the modular architecture shown in Figure 6.15. Let's describe what each of the components that appear in the figure are:

- **Map Server**: This component reads a map from two files and publishes it as a `nav_msgs/msg/OccupancyGrid`, which nodes internally handle as a `costmap2D`. The maps in Nav2 are grids whose cells encode whether the space is free (0), unknown (255), or occupied (254). Values between 1 and 253 encode different occupation degrees or cost to cross this area. Figure 6.16b shown the map coded as a `costmap2D`.

- **AMCL**: This component implements a localization algorithm based on Adaptive Monte-Carlo (AMCL) [6]. It uses sensory information, primarily dis-

---

[4]`https://navigation.ros.org`
[5]if using the Static Costmap Layer

 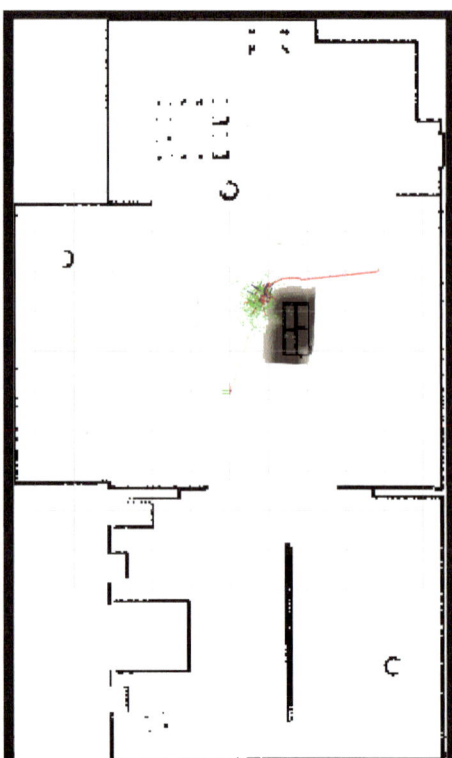

(a) Global costmap used by the Planner Server.

(b) Original map and the local costmap used by the Controller Server.

Figure 6.16: 2D costmaps used by the Nav2 components.

tance readings from a laser and the map, to calculate the robot's position. The output is a geometric transformation indicating the position of the robot. Since every frames should not have two parents, instead of posting a `map` → `base_footprint` transform, this component computes and publishes the `map` → `odom` transform.

- **Planner Server**: The function of this component is to calculate a route from the origin to the destination. It takes as input the destination, the current position of the robot, and the map of the environment. The Planner Server builds a costmap from the original map, whose walls are fattened with the radius of the robot and a certain safety margin. The idea is that the robot uses the free space (or with low cost) to calculate the routes, as shown in Figure 6.16a. Route planning and costmap updating algorithms are loaded as plugins. Like the following two components, this component receives requests through ROS 2 actions.

- **Controller Server**: This component receives the route calculated by the Planner Server and publishes the speeds sent to the robot base. It uses a costmap of the robot's surroundings (see Figure 6.16b), where nearby obstacles are encoded and used by algorithms (loaded as plugins) to calculate speeds.

- **Recovery Server**: This component has several helpful recovery strategies if the robot gets lost, gets stuck, or cannot calculate routes to the destination. These strategies are turning, clearing costmaps, slow-moving, among others.

- **BT Navigator Server**: This is the component that orchestrates the rest of the navigation components. It receives navigation requests in the form of ROS 2 actions. The action name is `navigate_to_pose` and the type is `nav2_msgs/action/NavigateToPose`. Therefore, if we want to make a robot go from one point to another, we must use this ROS 2 action. Check out what this action looks like:

```
ros2 interface show nav2_msgs/action/NavigateToPose

#goal definition
geometry_msgs/PoseStamped pose
string behavior_tree
---
#result definition
std_msgs/Empty result
---
geometry_msgs/PoseStamped current_pose
builtin_interfaces/Duration navigation_time
int16 number_of_recoveries
float32 distance_remaining
```

- The request section comprises a target position and, optionally, a custom Behavior Tree to be used in this action instead of the default one. This last feature allows special requests to be made that are not normal navigation behavior, such as following a moving object or approaching an obstacle in a particular way.
- The result of the action, when finished.
- The robot continuously returns the current position and the distance to the target and statistical data such as the navigation time or the times it has recovered from undesirable situations.

BT Navigator uses Behavior Trees to orchestrate robot navigation. The Behavior Tree nodes make requests to the other components of Nav2 so that they carry out their task.

When this component accepts a navigation action, it starts executing a Behavior Tree like the one shown in Figure 6.17. Nav2's default Behavior Tree is quite a bit more complex, including calls to recoveries, but the one in the figure is quite illustrative of BT Navigator Server's use of them. First, the goal that arrives in the ROS 2 action is put on the blackboard. ComputePathToPose uses this goal to call the Planner Server action, which returns a route to it. This path is the output of this BT node which is input to the BT node `FollowPath`, which sends it to the Controller Server.

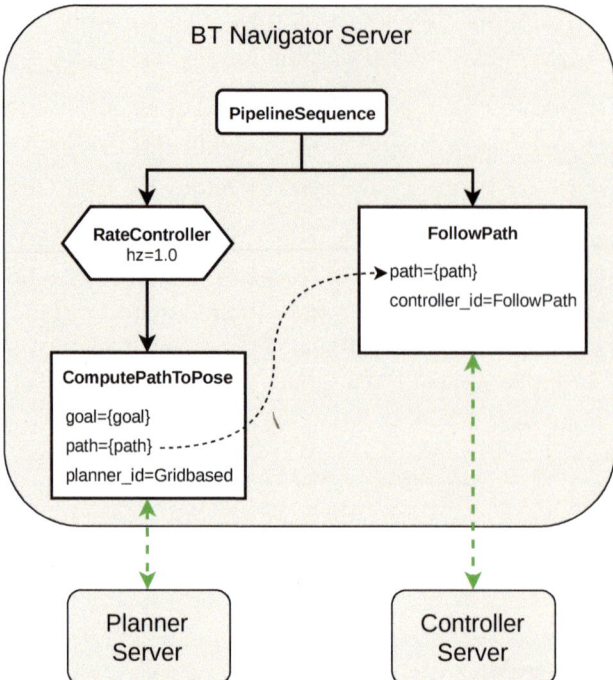

Figure 6.17: Behavior Tree simple example inside BT Navigator Server, with BT nodes calling ROS 2 actions to coordinate other Nav2 components.

To use Nav2, it is enough to install the packages that contain it:

```
$ sudo apt install ros-jazzy-navigation2 ros-jazzy-nav2-bringup
ros-jazzy-turtlebot3*
```

In the br2_navigation package, we have prepared the necessary launchers, maps, and configuration files for the simulated Tiago robot to navigate in the home scenario. Let's test navigation:

1. Launch the simulator:

```
$ ros2 launch br2_tiago sim.launch.py
```

2. Launch navigation:

```
$ ros2 launch br2_navigation tiago_navigation.launch.py
```

3. Open RViz2 and display (see Figure 6.18):

   - TF: To display the robot. Observe the transformation map → odom.
   - Map: Display the topic /map, which QoS is reliable and transient local.
   - Global Costmap: Display the topic /global_costmap/costmap with default QoS (Reliable and Volatile).

- Local Costmap: Display the topic `/local_costmap/costmap` with default QoS.
- LaserScan: To see how it matches with obstacles.
- It is interesting to display the AMCL particles. Each one is a hypothesis about the robot's position. The final robot position is the mean of all these particles. As much concentrated is this population of arrows, better localized is the robot. It is in the `/particlecloud` with which QoS is best effort + volatile.

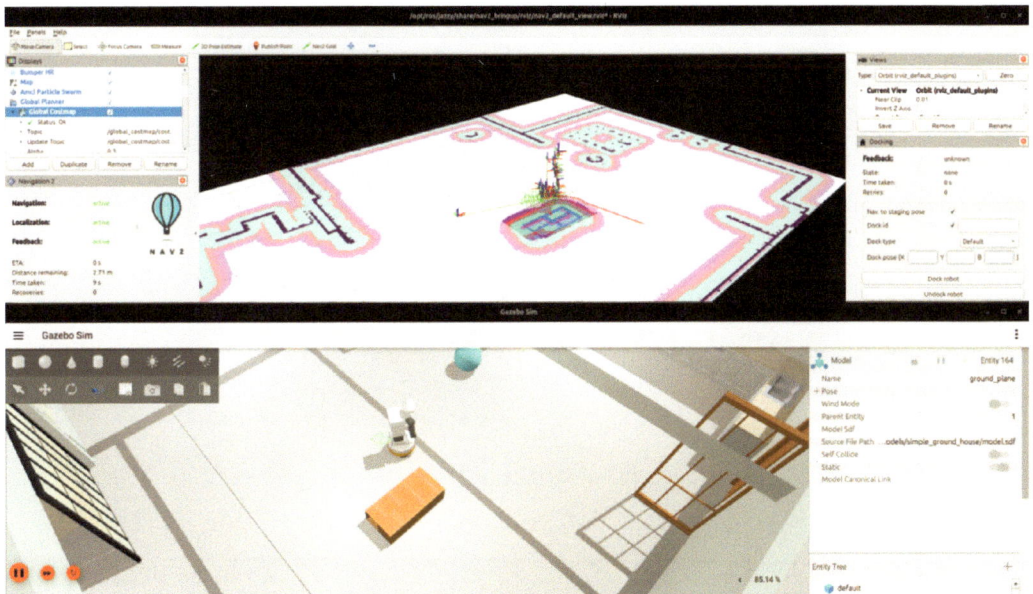

Figure 6.18: Nav2 in action

4. Use the "2D Goal Pose" button to command a goal position to the robot.

5. In obtaining a map position, use the "Publish Point" button. Then click in any position on the map. This position will be published to the topic `/clicked_point`.

6.3.2 Setup Nav2

This section describes the Nav2 setup process for a new environment and a specific robot. It is possible to can skip it, as the `br2_navigation` package already contains everything you need to make the simulated Tiago navigate in the house scenario. Keep reading for using another scenario or another robot.

If Nav2 is installed from packages, it is in `/opt/ros/jazzy/`. In particular, in `/opt/ros/jazzy/share/nav2_bringup` is the Nav2 bringup package with launchers, maps, and parameters for a simulated Turtlebot3[6] that comes by default and that you can launch by typing:

---

[6]`https://emanual.robotis.com/docs/en/platform/turtlebot3/overview`

```
$ ros2 launch nav2_bringup tb3_simulation_launch.py
```

It starts a simulation with a Turtlebot3 in a small world. Use the "2D Pose Estimate" button to put where the robot is (see Figure 6.19), as the navigation will not be activated until then.

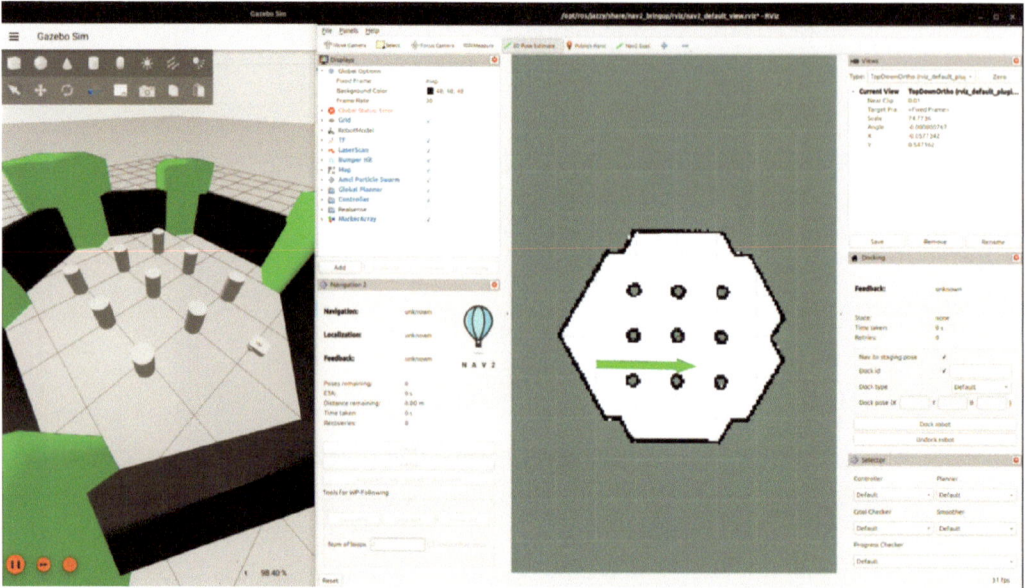

Figure 6.19: Simulated turtlebot 3.

The package for the Tiago simulation has been created copying some elements from nav2_bringup, since some extra remap in the launchers is needed, and thus having the configuration files and the maps together. This package has the following structure:

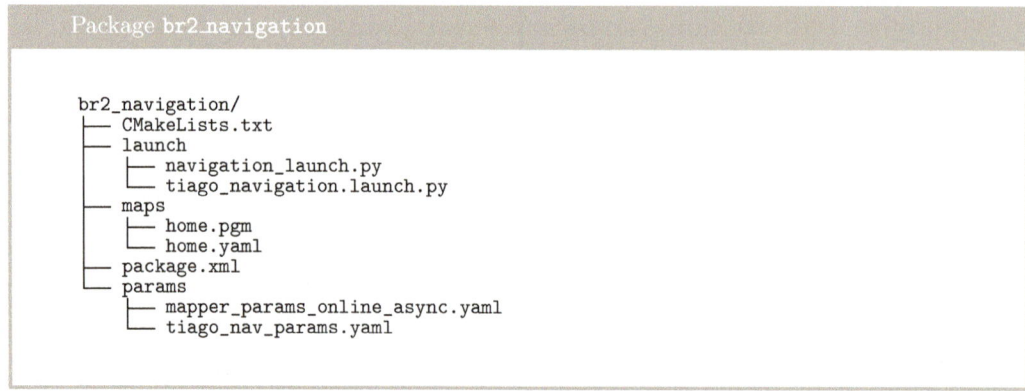

Package br2_navigation

```
br2_navigation/
├── CMakeLists.txt
├── launch
│   ├── navigation_launch.py
│   └── tiago_navigation.launch.py
├── maps
│   ├── home.pgm
│   └── home.yaml
├── package.xml
└── params
    ├── mapper_params_online_async.yaml
    └── tiago_nav_params.yaml
```

Start by looking at how to map the environment. We will use the slam_toolbox package. We will use a custom param file to specify the particular topics and frames:

```
params/mapper_params_online_async.yaml
```

```
# ROS Parameters
odom_frame: odom
map_frame: map
base_frame: base_footprint
scan_topic: /scan_raw
mode: mapping #localization
```

Run these commands, each in a different terminal:

1. Tiago simulated with the home scenario.

```
$ ros2 launch br2_tiago sim.launch.py
```

2. RViz2 to visualize the mapping progress (see Figure 6.20).

```
$ rviz2 --ros-args -p use_sim_time:=true
```

3. Launch the SLAM node. It will publish in /map the map as far as it is being built.

```
$ ros2 launch slam_toolbox online_async_launch.py params_file:=[Full path
to bookros2_ws/src/book_ros2/br2_navigation/params/mapper_params_online_async
.yaml] use_sim_time:=true
```

4. Launch the map saver server. This node will subscribe to /map, and it will save it to disk when requested.

```
$ ros2 launch nav2_map_server map_saver_server.launch.py
```

5. Run the teleoperator to move the robot along the scenario.

```
$ ros2 run teleop_twist_keyboard teleop_twist_keyboard --ros-args --remap
/cmd_vel:=/key_vel -p use_sim_time:=true
```

6.

Run these commands, each in a different terminal:

As soon as the robot starts moving around the stage using the teleoperator, run RViz2 and check how the map is built. When the map is completed, ask the map server saver to save the map to disk:

```
$ ros2 run nav2_map_server map_saver_cli --ros-args -p use_sim_time:=true
```

Note that when mapping/navigating with a real robot, the use_sim_time parameters, both in launchers and nodes, must be false.

At this point, two files will have been created. A PGM image file (which you can modify if you need to do any fix) and a YAML file containing enough information to interpret the image as a map. Remember that if modifying the name of the files, this YAML should be modified too:

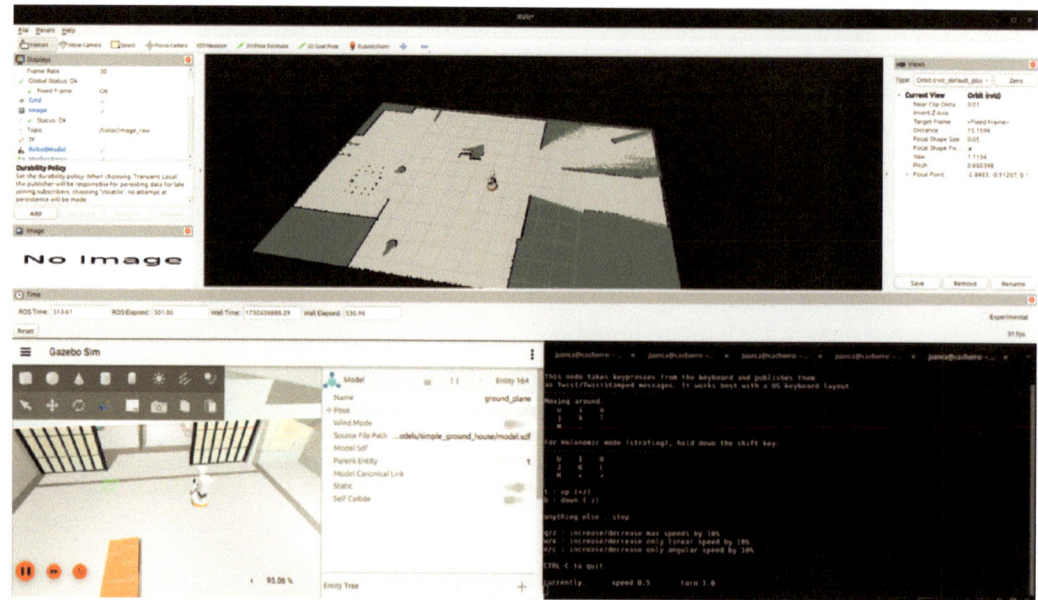

Figure 6.20: SLAM with Tiago simulated.

```
image: home.pgm
mode: trinary
resolution: 0.05
origin: [-2.46, -13.9, 0]
negate: 0
occupied_thresh: 0.65
free_thresh: 0.25
```

Move this file to the **br2_navigation** package and continue to the next setup step. In this step, the launchers copied from **nav2_bringup** needs to be modified. **tiago_navigation.launch** launch navigation and localization by including their launchers. We don't use directly the launchers in **nav2_bringup** because some extra remaps in **navigation.launch** has to be done.

```
br2_navigation/launch/navigation_launch.py

remappings = [('/tf', 'tf'),
('/tf_static', 'tf_static'),
('/cmd_vel', '/nav_vel')
]
```

Regarding the parameter files, start from the ones in the package **nav2_bringup**. Let's see some details on the configuration:

- First, and most important, set all the parameters that contain a sensor topic to the correct ones, and ensure that all the frames exist in our robot and are correct.

- If the initial position is known, set it in the AMCL configuration. If you start the robot in the same pose as you started when mapping, this is the $(0, 0, 0)$ position.

```
br2_navigation/params/tiago_nav_params

amcl:
  ros__parameters:
    scan_topic: scan_raw
    set_initial_pose: true
    initial_pose:
    x: 0.0
    y: 0.0
    z: 0.0
    yaw: 0.0
```

- Set the speeds and acceleration depending on the robot's capabilities:

```
br2_navigation/params/tiago_nav_params

controller_server:
  ros__parameters:
    use_sim_time: False
    FollowPath:
      plugin: "dwb_core::DWBLocalPlanner"
      min_vel_x: 0.0
      min_vel_y: 0.0
      max_vel_x: 0.3
      max_vel_y: 0.0
      max_vel_theta: 0.5
      min_speed_xy: 0.0
      max_speed_xy: 0.5
      min_speed_theta: 0.0
      acc_lim_x: 1.5
      acc_lim_y: 0.0
      acc_lim_theta: 2.2
      decel_lim_x: -2.5
      decel_lim_y: 0.0
      decel_lim_theta: -3.2
```

- Set the robot radius to inflate walls and obstacles and a scaling factor in setting how far navigate from them. These settings are held by the `inflation_layer` costmap plugin, applicable to local and global costmap:

```
br2_navigation/params/tiago_nav_params

local_costmap:
  local_costmap:
    ros__parameters:
      robot_radius: 0.3
      plugins: ["voxel_layer", "inflation_layer"]
      inflation_layer:
        plugin: "nav2_costmap_2d::InflationLayer"
        cost_scaling_factor: 3.0
        inflation_radius: 0.55
```

6.3.3 Computation Graph and Behavior Tree

The Computation Graph (Figure 6.21) of this project is made up of the node `patrolling_node` and the nodes that belong to the two subsystems that are being controlled: Nav2 and the active vision system developed in the last chapter.

- Nav2 is controlled using ROS 2 actions, sending the goal poses that make up the patrol route.

- Regarding the active vision system during navigation, the HeadController node, a LifeCycleNode, will be activated (using ROS 2 services).

- Also, upon arrival at a waypoint, to make the robot rotate on its own, `patrolling_node` will post velocities directly to the robot's base.

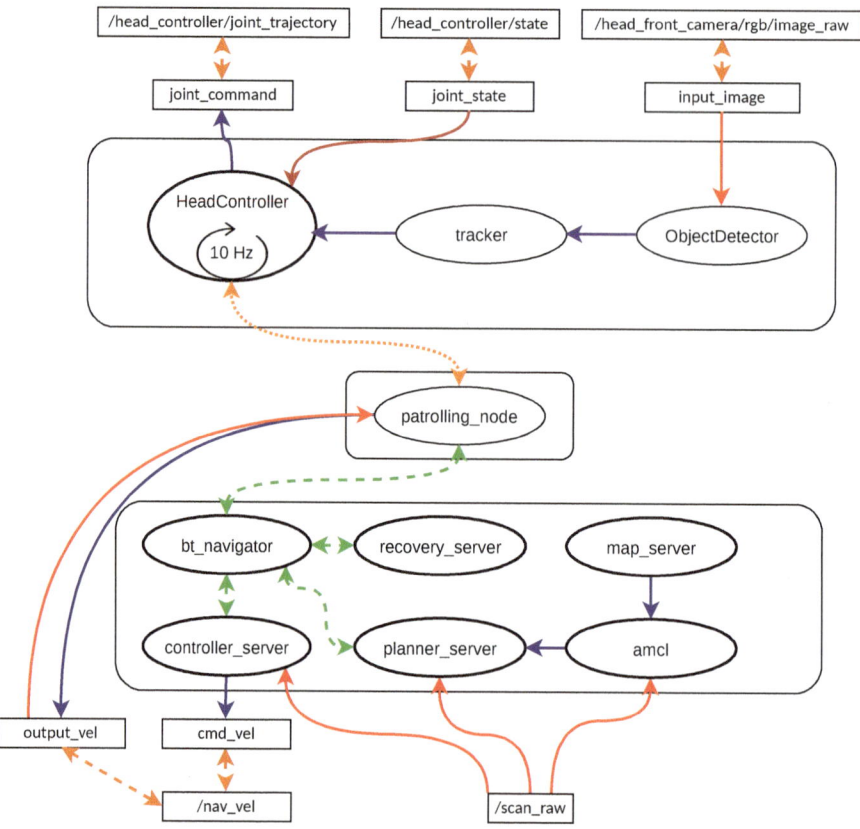

Figure 6.21: Computation Graph for the Patrolling Project. Subsystems have been simplified for clarity.

The `patrolling_node` node in the Computation Graph is shown to be quite simple. Perhaps it is more interesting to analyze the Behavior Tree that it contains, which is the one that controls its control logic. Figure 6.22 shows its complete structure. Analyze each one of its action and condition nodes:

- **Move**: This node is in charge of sending a navigation request to Nav2 through a ROS 2 action. The navigation goal is received through an input port, in its **goal** port, which is a coordinate that contains an (x, y) position and a *theta* orientation. This node returns RUNNING until it is informed that the navigation action is complete, in which case it returns SUCCESS. The case in which it returns FAILURE has not been contemplated, although it would have been convenient.

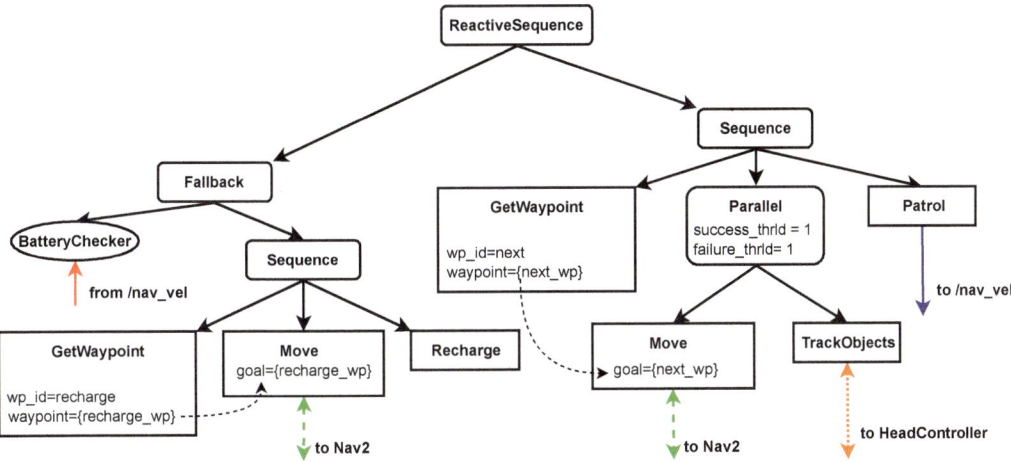

Figure 6.22: Behavior Tree for Patrolling project.

- **GetWaypoint**: This node is used to obtain the geometric coordinates used by Move. It has a waypoint output port with the geometric coordinates, which are then used by the BT node as input. The GetWayPoint input is an id indicating which waypoint is desired. If this input is "recharge", its output is the coordinates of the recharge point. If the input is "next", it returns the geometric coordinates of the next waypoint to navigate.

  This node exists because it simplifies the Behavior Tree since otherwise, the right branch of the tree would have to be repeated three times, once for each waypoint. The second is to delegate to another BT node the choice of the target point and thus simplify Move, not needing to maintain the coordinates of all the waypoints internally. There are many more alternatives, but this one is pretty clean and scalable.

- **BatteryChecker**: This node simulates the battery level of the robot. It keeps the battery level on the blackboard, decreasing over time and with the robot's movement (that is why it subscribes to the topic of commanded speeds). If the battery level drops below a certain level, it returns FAILURE. If not, return SUCCESS.

- **Patrol**: This node simply spins the robot around for a few seconds to control the environment. When it has finished, it returns SUCCESS.

- **TrackObjects**: This node always returns RUNNING. When it is first ticked, it activates, if it was not already, the HeadController node. This node runs in parallel with Move. The Parallel control node is configured so that when one of the two (and it can only be Move) returns SUCCESS, it considers that the task of all its children has finished, halting the nodes whose status is still RUNNING. When TrackObjects receives a halt, it disables the HeadController.

6.3.4 Patrolling Implementation

The structure of the `br2_bt_patrolling` package is similar to the one in the previous section: it implements each BT node separately, in the usual places for class definitions and declarations. It has a main program that creates the tree and executes it, and it has some tests for each of the implemented BT nodes.

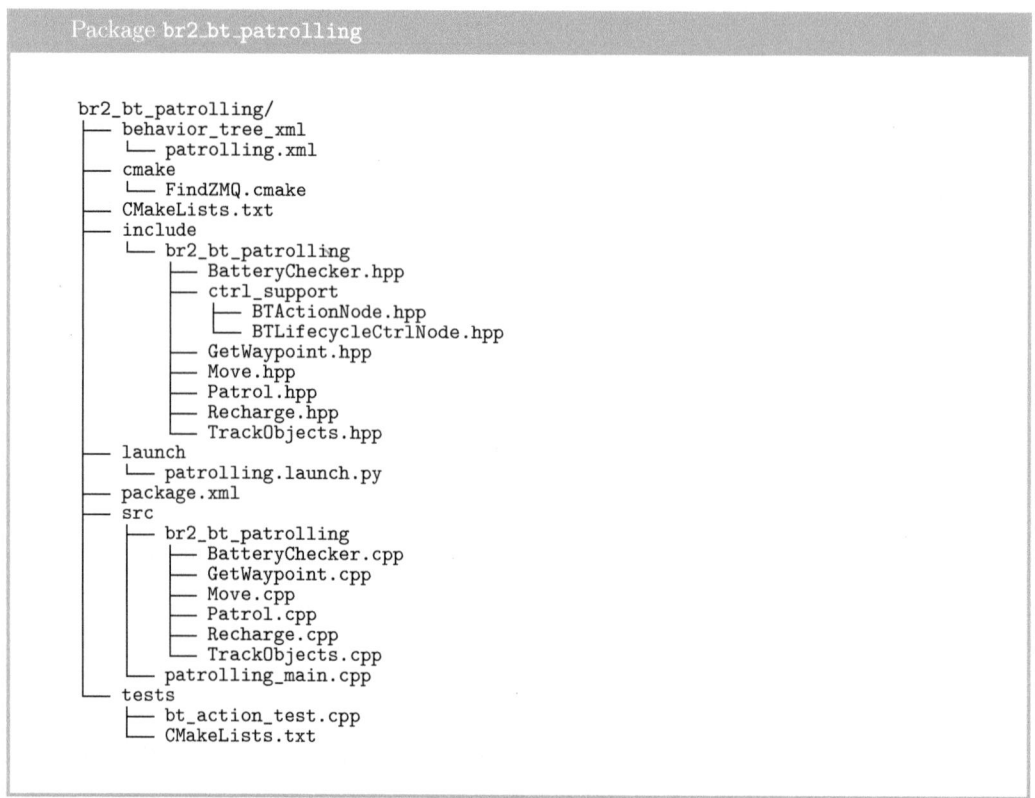

```
Package br2_bt_patrolling


  br2_bt_patrolling/
  ├── behavior_tree_xml
  │   └── patrolling.xml
  ├── cmake
  │   └── FindZMQ.cmake
  ├── CMakeLists.txt
  ├── include
  │   └── br2_bt_patrolling
  │       ├── BatteryChecker.hpp
  │       ├── ctrl_support
  │       │   ├── BTActionNode.hpp
  │       │   └── BTLifecycleCtrlNode.hpp
  │       ├── GetWaypoint.hpp
  │       ├── Move.hpp
  │       ├── Patrol.hpp
  │       ├── Recharge.hpp
  │       └── TrackObjects.hpp
  ├── launch
  │   └── patrolling.launch.py
  ├── package.xml
  ├── src
  │   ├── br2_bt_patrolling
  │   │   ├── BatteryChecker.cpp
  │   │   ├── GetWaypoint.cpp
  │   │   ├── Move.cpp
  │   │   ├── Patrol.cpp
  │   │   ├── Recharge.cpp
  │   │   └── TrackObjects.cpp
  │   └── patrolling_main.cpp
  └── tests
      ├── bt_action_test.cpp
      └── CMakeLists.txt
```

From an implementation point of view, the most interesting are two classes that simplify the BT nodes that use ROS 2 actions and those that activate a LifeCycleNode. They are in `include/br2_bt_patrolling/ctrl_support`, and have been implemented in a general way so that they can be reused for other projects.

The `BTActionNode` class has been borrowed from Nav2, where the BT Navigator Server used it to control the rest of its servers. It is quite a complex class since it considers many more cases than we use in this project, such as cancellation and resends of actions. We do not want to go into details about its implementation. I recommend the ROS 2 actions tutorial on the official ROS 2 page to learn more about ROS 2 actions. When completed, come back to this class to explore this class.

BT nodes that wish to control a subsystem with ROS 2 actions inherit this class. Let's analyze its interface to its derived class. Original comments will help us to understand their utility:

```
include/br2_bt_patrolling/ctrl_support/BTActionNode.hpp
```

```cpp
template<class ActionT, class NodeT = rclcpp::Node>
class BtActionNode : public BT::ActionNodeBase
{
public:
  BtActionNode(
    const std::string & xml_tag_name,
    const std::string & action_name,
    const BT::NodeConfiguration & conf)
  : BT::ActionNodeBase(xml_tag_name, conf), action_name_(action_name)
  {
    node_ = config().blackboard->get<typename NodeT::SharedPtr>("node");
    ...
  }

  // Could do dynamic checks, such as getting updates to values on the blackboard
  virtual void on_tick()
  {
  }

  // Called upon successful completion of the action. A derived class can override this
  // method to put a value on the blackboard, for example.
  virtual BT::NodeStatus on_success()
  {
    return BT::NodeStatus::SUCCESS;
  }

  // Called when a the action is aborted. By default, the node will return FAILURE.
  // The user may override it to return another value, instead.
  virtual BT::NodeStatus on_aborted()
  {
    return BT::NodeStatus::FAILURE;
  }

  // The main override required by a BT action
  BT::NodeStatus tick() override
  {
    ...
  }

  // The other (optional) override required by a BT action. In this case, we
  // make sure to cancel the ROS 2 action if it is still running.
  void halt() override
  {
    ...
  }
protected:
  typename ActionT::Goal goal_;
};
```

It is a template class because each action has a different type. In the case of Move, the action type is nav2_msgs/action/NavigateToPose. The class is also parameterized with the ROS 2 node type because it may also be instantiated with LifeCycleNodes.

The tick and halt methods are handled by class BtActionNode, so they should not be defined in the derived class. The other methods can be overridden in the derived class to do something, like notifying when the action completes or fails. The derived class overrides on_tick, which is called once at startup, to set the goal. Let's see the implementation of Move inheriting from BtActionNode:

```
include/br2_bt_patrolling/Move.hpp

class Move : public br2_bt_patrolling::BtActionNode<nav2_msgs::action::NavigateToPose>
{
public:
  explicit Move(
    const std::string & xml_tag_name,
    const std::string & action_name,
    const BT::NodeConfiguration & conf);

  void on_tick() override;
  BT::NodeStatus on_success() override;

  static BT::PortsList providedPorts()
  {
    return {
      BT::InputPort<geometry_msgs::msg::PoseStamped>("goal")
    };
  }
};
```

```
src/br2_bt_patrolling/Move.cpp

Move::Move(
  const std::string & xml_tag_name,
  const std::string & action_name,
  const BT::NodeConfiguration & conf)
: br2_bt_patrolling::BtActionNode<nav2_msgs::action::NavigateToPose>(xml_tag_name,
    action_name, conf)
{
}

void
Move::on_tick()
{
  geometry_msgs::msg::PoseStamped goal;
  getInput("goal", goal);

  goal_.pose = goal;
}

BT::NodeStatus
Move::on_success()
{
  RCLCPP_INFO(node_->get_logger(), "navigation Suceeded");

  return BT::NodeStatus::SUCCESS;
}

#include "behaviortree_cpp_v3/bt_factory.h"
BT_REGISTER_NODES(factory)
{
  BT::NodeBuilder builder =
    [](const std::string & name, const BT::NodeConfiguration & config)
    {
      return std::make_unique<br2_bt_patrolling::Move>(
        name, "navigate_to_pose", config);
    };

  factory.registerBuilder<br2_bt_patrolling::Move>(
    "Move", builder);
}
```

- The BT Node Move implements on_success to report that the navigation has finished.

- The on_tick method gets the goal from the input port and assigns it to goal_. This variable will be the one that will be sent directly to Nav2.

- When building this BT node, the second argument is the name of the ROS 2 action. In the case of Nav2, it is `navigate_to_pose`.

`BTLifecycleCtrlNode` is a class from which a BT Node is derived to activate/deactivate LifeCycle nodes. It is created specifying the name of the node to control. In the case of the `HeadTracker`, it will be with `/head_tracker`. All LifeCycleNodes have various services to be managed. In this, we will be interested in two:

- `[node name]/get_state`: Returns the status of a LifeCycleNode.

- `[node name]/set_state`: Sets the state of a LifeCycleNode.

Let's see code snippets of the `BTLifecycleCtrlNode` implementation:

```cpp
include/br2_bt_patrolling/ctrl_support/BTLifecycleCtrlNode.hpp

class BtLifecycleCtrlNode : public BT::ActionNodeBase
{
public:
  BtLifecycleCtrlNode(...)
  : BT::ActionNodeBase(xml_tag_name, conf), ctrl_node_name_(node_name)
  {
  }

  template<typename serviceT>
  typename rclcpp::Client<serviceT>::SharedPtr createServiceClient(
    const std::string & service_name)
  {
    auto srv = node_->create_client<serviceT>(service_name);
    while (!srv->wait_for_service(1s)) {
      ...
    }
    return srv;
  }

  BT::NodeStatus tick() override
  {
    if (status() == BT::NodeStatus::IDLE) {
      change_state_client_ = createServiceClient<lifecycle_msgs::srv::ChangeState>(
        ctrl_node_name_ + "/change_state");
      get_state_client_ = createServiceClient<lifecycle_msgs::srv::GetState>(
        ctrl_node_name_ + "/get_state");
    }

    if (ctrl_node_state_ != lifecycle_msgs::msg::State::PRIMARY_STATE_ACTIVE) {
      ctrl_node_state_ = get_state();
      set_state(lifecycle_msgs::msg::State::PRIMARY_STATE_ACTIVE);
    }

    return BT::NodeStatus::RUNNING;
  }

  void halt() override
  {
    if (ctrl_node_state_ == lifecycle_msgs::msg::State::PRIMARY_STATE_ACTIVE) {
      set_state(lifecycle_msgs::msg::State::PRIMARY_STATE_INACTIVE);
    }
  }

  // Get the state of the controlled node
  uint8_t get_state(){...}

  // Set the state of the controlled node. It can fail, if no transition is possible
  bool set_state(uint8_t state) {...}

  std::string ctrl_node_name_;
  uint8_t ctrl_node_state_;
};
```

Two clients are instantiated: one to query the state and one to set the state. They will be used in `get_state` and `set_state`, respectively. When the node is first ticked, the controlled node is requested to go to the active state. When halted, its deactivation is requested.

The BT node TrackObjects only needs to inherit from this class by specifying the name of the node:

```
include/br2_bt_patrolling/TrackObjects.hpp


class TrackObjects : public br2_bt_patrolling::BtLifecycleCtrlNode
{
public:
  explicit TrackObjects(
    const std::string & xml_tag_name,
    const std::string & node_name,
    const BT::NodeConfiguration & conf);

  static BT::PortsList providedPorts()
  {
    return BT::PortsList({});
  }
};
```

```
src/br2_bt_patrolling/TrackObjects.cpp


TrackObjects::TrackObjects(...)
: br2_bt_patrolling::BtLifecycleCtrlNode(xml_tag_name, action_name, conf)
{
}

#include "behaviortree_cpp_v3/bt_factory.h"
BT_REGISTER_NODES(factory)
{
  BT::NodeBuilder builder =
    [](const std::string & name, const BT::NodeConfiguration & config)
    {
      return std::make_unique<br2_bt_patrolling::TrackObjects>(
        name, "/head_tracker", config);
    };

  factory.registerBuilder<br2_bt_patrolling::TrackObjects>(
    "TrackObjects", builder);
}
```

Take into account that `TrackObjects` always returns RUNNING. That is why we have used it as a child of a parallel control node.

Check out how the rest of the BT nodes functionality has been implemented:

- **BatteryChecker**: The first difference between this BT node and the others is that it is a condition node. It does not have a `halt` method and cannot return RUNNING.

  This node checks the battery level stored on the blackboard at each tick. If it is less than a certain level, it returns FAILURE.

```
src/br2_bt_patrolling/BatteryChecker.cpp

    const float MIN_LEVEL = 10.0;

BT::NodeStatus
BatteryChecker::tick()
{
  update_battery();

  float battery_level;
  config().blackboard->get("battery_level", battery_level);
  if (battery_level < MIN_LEVEL) {
    return BT::NodeStatus::FAILURE;
  } else {
    return BT::NodeStatus::SUCCESS;
  }
}
```

The `update_battery` method takes the battery level from the blackboard and decreases it in the function of time and the total amount of speed (`last_twist_`) currently requested. It is just a simulation of battery consumption.

```
src/br2_bt_patrolling/BatteryChecker.cpp

    const float DECAY_LEVEL = 0.5;  // 0.5 * |vel| * dt
    const float EPSILON = 0.01;  // 0.01 * dt

    void
    BatteryChecker::update_battery()
    {
      float battery_level;
      if (!config().blackboard->get("battery_level", battery_level)) {
        battery_level = 100.0f;
      }
      float dt = (node_->now() - last_reading_time_).seconds();
      last_reading_time_ = node_->now();

      float vel = sqrt(last_twist_.linear.x * last_twist_.linear.x +
        last_twist_.angular.z * last_twist_.angular.z);
      battery_level = std::max(
        0.0f, battery_level - (vel * dt * DECAY_LEVEL) - EPSILON * dt);
      config().blackboard->set("battery_level", battery_level);
    }
```

It is always useful to control de range of some calculus using `std::max` and `std::min`. In this case, we control that `battery_level` is never negative.

- **Recharge**: This BT node is related to the previous one. It takes some time to recharge the battery. Note that using blackboard lets some nodes collaborate to update and test some values.

```
src/br2_bt_patrolling/BatteryChecker.cpp

    BT::NodeStatus
    Recharge::tick()
    {
      if (counter_++ < 50) {
        return BT::NodeStatus::RUNNING;
      } else {
        counter_ = 0;
        config().blackboard->set<float>("battery_level", 100.0f);
        return BT::NodeStatus::SUCCESS;
      }
    }
```

Each BT node in the tree is a different instance of the same class, but be ready to be ticked even if the BT node returned once SUCCESS. In this case, restarting `counter_` to 0.

- **Patrol**: This node just makes the robot spin for 15 s. The only interesting aspect of this node is how it controls how long executing since the first tick until the it return SUCCESS. Take into account that a node status is IDLE the first tick, so it's possible to store this timestamp.

```
src/br2_bt_patrolling/Patrol.cpp

BT::NodeStatus
Patrol::tick()
{
  if (status() == BT::NodeStatus::IDLE) {
    start_time_ = node_->now();
  }

  geometry_msgs::msg::Twist vel_msgs;
  vel_msgs.angular.z = 0.5;
  vel_pub_->publish(vel_msgs);

  auto elapsed = node_->now() - start_time_;

  if (elapsed < 15s) {
    return BT::NodeStatus::RUNNING;
  } else {
    return BT::NodeStatus::SUCCESS;
  }
}
```

- **GetWaypoint**: This node stores the waypoint coordinates. If the input port `wp_id` is the string "recharge", its output is a coordinate, in frame `map`, corresponding to the position where it is supposed to be the robot charger. In another case, each time it is ticked, it returns the coordinates of a different waypoint.

```
src/br2_bt_patrolling/GetWaypoint.cpp

GetWaypoint::GetWaypoint(...)
{
  geometry_msgs::msg::PoseStamped wp;
  wp.header.frame_id = "map";
  wp.pose.orientation.w = 1.0;

  // recharge wp
  wp.pose.position.x = 3.67;
  wp.pose.position.y = -0.24;
  recharge_point_ = wp;

  // wp1
  wp.pose.position.x = 1.07;
  wp.pose.position.y = -12.38;
  waypoints_.push_back(wp);

  // wp2
  wp.pose.position.x = -5.32;
  wp.pose.position.y = -8.85;
  waypoints_.push_back(wp);
}
```

```
src/br2_bt_patrolling/GetWaypoint.cpp

BT::NodeStatus
GetWaypoint::tick()
{
  std::string id;
  getInput("wp_id", id);

  if (id == "recharge") {
    setOutput("waypoint", recharge_point_);
  } else {
    setOutput("waypoint", waypoints_[current_++]);
    current_ = current_ % waypoints_.size();
  }

  return BT::NodeStatus::SUCCESS;
}
```

6.3.5 Running Patrolling

From an implementation point of view, the only relevant thing is that we will use a launcher for the active vision system and the patrolling node. Navigation and the simulator could have been included in the launcher, but they generate so much output on the screen that we run them manually in other terminals. The launcher looks like this:

```
br2_navigation/launch/patrolling_launch.py

def generate_launch_description():
    tracking_dir = get_package_share_directory('br2_tracking')

    tracking_cmd = IncludeLaunchDescription(
        PythonLaunchDescriptionSource(os.path.join(tracking_dir, 'launch',
                                                   'tracking.launch.py')))

    patrolling_cmd = Node(
        package='br2_bt_patrolling',
        executable='patrolling_main',
        parameters=[{
          'use_sim_time': True
        }],
        remappings=[
          ('input_scan', '/scan_raw'),
          ('output_vel', '/nav_vel')
        ],
        output='screen'
    )

    ld = LaunchDescription()
    ld.add_action(tracking_cmd)
    ld.add_action(patrolling_cmd)

    return ld
```

So type these commands, each one in a separate terminal:

```
$ ros2 launch br2_tiago sim.launch.py
```

```
$ ros2 launch br2_navigation tiago_navigation.launch.py
```

Nav2 also uses Behavior Trees and activates a server to debug its operation with Groot. It does this on Groot's default ports (1666 and 1667). For this reason, we have started it in 2666 and 2667. If we put them on the same, the program would fail.

Before connecting to the patrolling Behavior Tree, correctly set the ports to 2666 and 2667.

```
src/patrolling_main.cpp

    BT::Tree tree = factory.createTreeFromFile(xml_file, blackboard);
    auto publisher_zmq = std::make_shared<BT::PublisherZMQ>(tree, 10, 2666, 2667);
```

At this point, optionally open RViz2 to monitor navigation or Groot to monitor Behavior Tree execution. For the latter, wait to launch the patrol program to connect to the Behavior Tree:

```
$ rviz2 --ros-args -p use_sim_time:=true
```

Try sending a navigation position to make sure the navigation starts correctly.

```
$ ros2 run groot Groot
```

Finally, launch the patrol program together with the active vision system:

```
$ ros2 launch br2_bt_patrolling patrolling.launch.py
```

If everything has gone well, the robot, after recharging its battery, patrols the three waypoints established in the environment. While patrolling, observe how the robot tracks the objects it detects. When it reaches a waypoint and turns around, notice how tracking is no longer active. After a while of operation, the robot will run out of battery again, going to the recharging point again before continuing patrolling.

PROPOSED EXERCISES:

1. Make a program using Behavior Trees that makes the robot move continuously to the space without obstacles.

2. Explore the Nav2 functionality:

   • Mark forbidden areas in the center of each room in which the robot should not enter.

   • Modify the Behavior tree inside BT Navigator to finish navigation always one meter before the goal.

   • Try different Controller/Planner algorithms.

3. Publish the detected objects while patrolling as a 3D bounding box. You could do it by:

   • Using the pointcloud.

   • Using the depth image and the CameraInfo information. Like is done in:

     https://github.com/gentlebots/gb_perception/blob/main/gb_perception_utils/src/gb_perception_utils/Perceptor3D.cpp

Deep ROS 2

S o far, we have covered the basic concepts that a beginner in ROS 2 should understand to develop robot applications using ROS 2. This chapter aims to go beyond that, explaining the internal workings of nodes and some of ROS's core mechanisms. This deeper dive will reveal what is happening "below the surface", allowing us to better understand certain effects that may occur during node execution or apply real-time programming principles, which will be discussed toward the end of the chapter.

7.1 ROS 2 EXECUTION MANAGEMENT

The basic unit of execution in ROS 2 is the Node. A ROS 2 application is a computational graph where nodes collaborate to accomplish the task they were programmed for. Nodes run within system processes, and multiple nodes can operate within the same process. A node is programmed following a model where the logic resides in the callbacks triggered by events: timer callbacks for code that needs to run periodically, and subscriber/service callbacks for handling incoming data or requests directed to the node.

The entities responsible for managing these events and invoking the callbacks are the executors, and they play a crucial role in execution management within ROS 2. Executors are always present, even when we might not explicitly think about them. For example, in the following code, there appears to be no executor at all:

```cpp
int main(int argc, char * argv[]) {
  rclcpp::init(argc, argv);

  auto node = rclcpp::Node::make_shared("listener");
  auto sub = node->create_subscription<std_msgs::msg::String>(
    "/chatter", callback);

  rclcpp::spin(node);
  rclcpp::shutdown();
}
```

In reality, it uses a **SingleThreadedExecutor**, as the implementation of `rclcpp::spin` is as follows:

DOI: 10.1201/9781003516798-7

ros2/rclcpp/rclcpp/src/rclcpp/executors.cpp

```cpp
void
rclcpp::spin(rclcpp::node_interfaces::NodeBaseInterface::SharedPtr node_ptr)
{
  rclcpp::ExecutorOptions options;
  options.context = node_ptr->get_context();
  rclcpp::executors::SingleThreadedExecutor exec(options);
  exec.add_node(node_ptr);
  exec.spin();
  exec.remove_node(node_ptr);
}
```

It is important to understand at which layer each of the elements involved in the execution management of a node operates. To illustrate the following explanation, I have created Figure 7.1, which on the right contains a simplified version of Figure 1.7, and on the left shows the relevant elements discussed in the explanation, which are:

- At the bottom, within the middleware layer, are the topics, where we have represented incoming messages to each topic with circles, along with the order in which they arrived at the node. These messages are stored in queues within each middleware.

- The user node layer contains the callback code that must be executed to process the messages located in the middleware layer.

- In the client libraries layer (rclcpp, rclpy,...), we find the executors. Executors are responsible for transferring messages from the middleware layer to the callbacks (cb_1, cb_2, and cb_3) in the user layer. How this is done has a significant impact on the execution model in ROS 2.

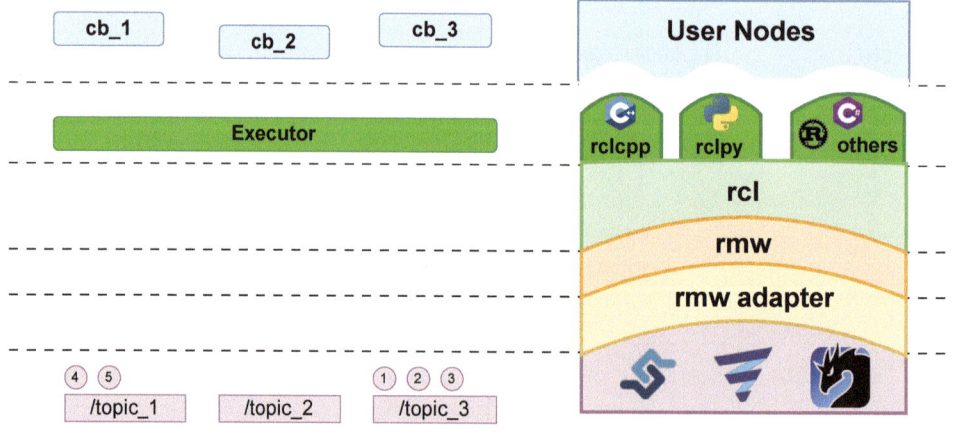

Figure 7.1: Execution management in ROS 2 layers.

To process the messages arriving at the node, an executor performs several steps, as shown in Figure 7.2:

1. An executor uses a wait-set mechanism to wait for a message to arrive in the queue of one of the subscribers from the nodes registered with it. A wait-set is a mechanism that allows a thread (or a set of threads) to wait for one or more specific events to occur, such as the presence of messages in the queue.

   The wait-set acts as a "waiting set" where the conditions or events that a thread is waiting for are registered, and the thread is suspended until one of these events occurs. When any of the registered events is triggered, the thread is notified to resume execution. A wait-set optimizes system resource usage by allowing threads to wake up only when there is actual work to be done, instead of continuously checking for new events (a technique known as *busy waiting*, which is less efficient).

   As soon as the wait-set is activated, the subscribers with messages waiting in their queues are marked in the wait-set conditions.

2. Once an executor knows which queues have messages to process, for each of the queues with messages, it retrieves one message and,

3. executes the callback associated with the subscriber of that queue, providing the retrieved message.

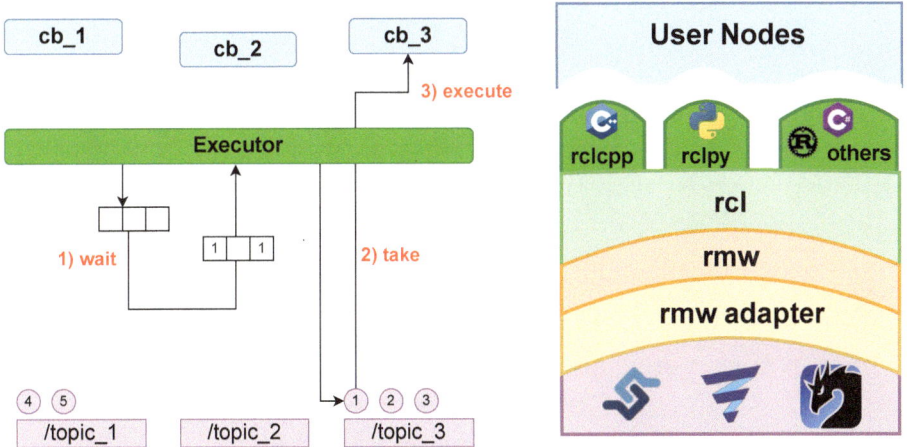

Figure 7.2: Steps involved in processing a message by an executor.

Executors perform this task using one or more threads, depending on whether it is a `SingleThreadedExecutor` or a `MultiThreadedExecutor`. How the executor is implemented defines its semantics—that is, how it waits for events, processes pending events, and adapts to changes in the nodes.

The executors mentioned earlier follow a semantic flow, illustrated in Figure 7.3, and described as follows:

1. The executor retrieves information from the wait-set, provided by the middleware, indicating if there are incoming messages or other events ready to be processed.

2. If there is a timer event in the wait-set, the associated callback is executed, the event is removed from the wait-set, and the process returns to step 1.

3. If a topic in the wait-set needs to be processed, the callback associated with this message is invoked, the topic is removed from the wait-set, and the process returns to step 1.

4. For a service server, if there is a service request in the wait-set, the associated callback is executed, the service is removed from the wait-set, and the process returns to step 1.

5. For a service client, if there is a response to a request from the service server, the associated callback is executed, the service is removed from the wait-set, and the process returns to step 1.

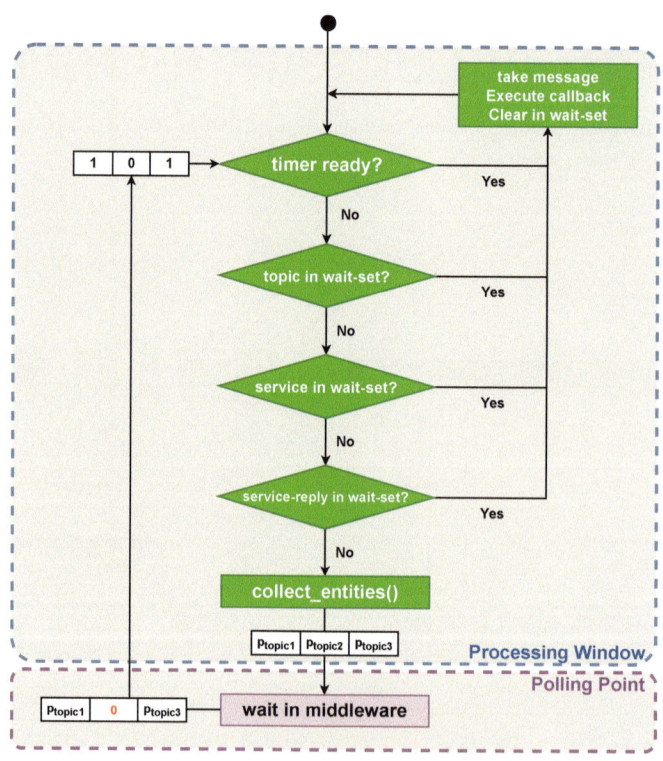

Figure 7.3: Executor Semantic.

6. Once the wait-set has been processed, the `collect_entities()` function gathers all timers, subscriptions, and services from the nodes[1] within the executor and creates a new wait-set vector, which is sent to the middleware. This process informs the middleware which elements we are interested in being notified about when they have new messages or events.

---

[1] as we will soon see, actually from something called a *callback group*

7. The middleware waits for new messages or events in the provided wait-set, fills the wait-set with those that have messages or events, and the process returns to step 1.

Figure 7.4 shows the situation from Figure 7.2 where three messages arrived at the topic /topic_1 and then two more at /topic_2. The wait-set after the Polling Point indicates that there is data in both topics, so during this processing window, the messages in both topics are processed in FIFO order. This results in message 1 and message 4 being processed. Note that although messages 2 and 3 arrived before message 4, they will be processed after the next Polling Point, when messages 2 and 5 will be handled, and in a subsequent Polling Point, message 3 will be processed.

By default, callbacks and events within the same node are executed sequentially, even if we are using a MultiThreadedExecutor. This ensures that there will be no race conditions between them. We can only take advantage of a MultiThreadedExecutor, unless explicitly configured otherwise (which we will cover shortly), by processing messages or events from different nodes.

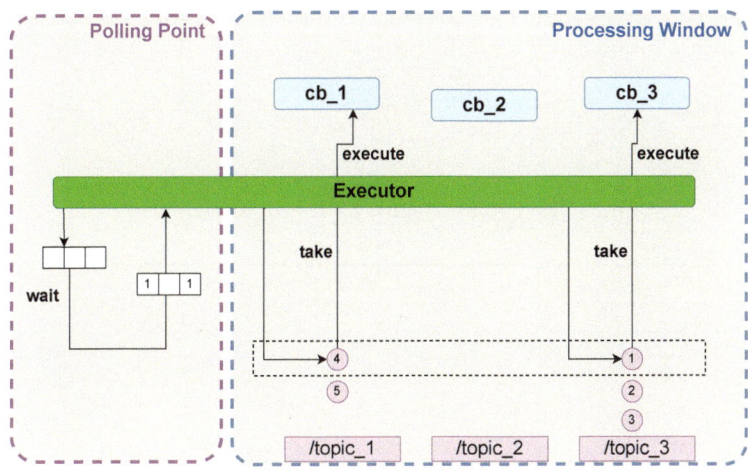

Figure 7.4: Execution of callbacks from a wait-set.

Another important detail to keep in mind is that the order in which messages are processed depends on the order in which they were created within the node. If the subscriber to /topic_1 was created before the one for /topic_2, the polling point will first call the callback for /topic_2 and then for /topic_1. If we want the messages from one topic to be prioritized over others when both have messages, we should create the preferred topic last. In many applications, this does not make a significant difference, but in critical applications, this processing order can have an impact.

In section 2.3.5, we introduced the executors that are already familiar to us: the SingleThreadedExecutor and the MultiThreadedExecutor. At this point, I would like to introduce two new executors:

- **StaticSingleThreadedExecutor**: This executor does not rescan nodes to identify new subscribers or timers, but only does so at the start of execution, as shown in Figure 7.5, saving the computation time that would otherwise be spent scanning at each polling point. The downside is that all subscribers, timers, and services must be created at the start, because if they appear later, their messages or events will not be considered.

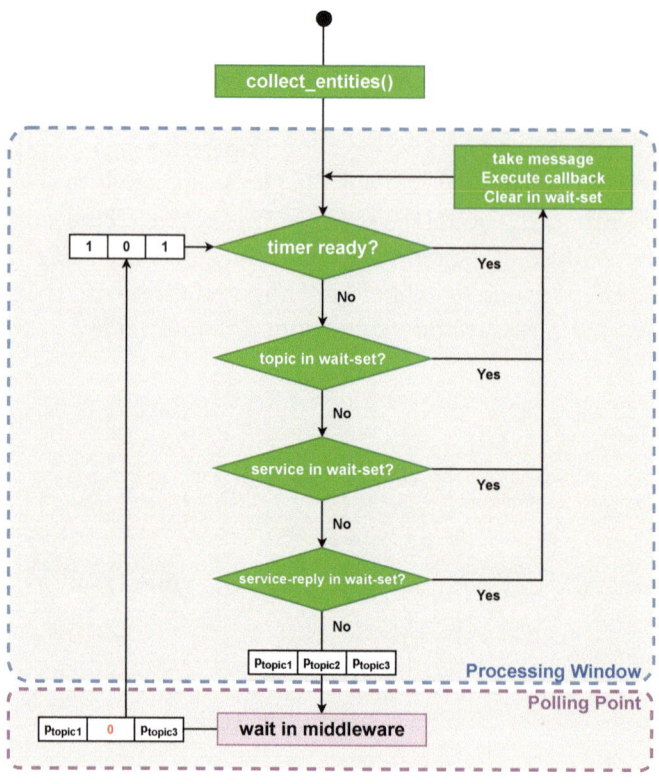

Figure 7.5: Executor Semantic for StaticSingleThreadedExecutor.

- **EventsExecutor**: This recently added executor in rclcpp does not use a wait-set, but instead leverages the RMW listener API to queue all events as they occur and process them in that order. Unlike other executors, timer management no longer goes through the middleware. This executor, still experimental at the time this book was written, has shown experimental time savings as it eliminates the overhead caused by the wait-set-based strategy. The main limitation is that this executor operates with only a single thread.

The decision of which executor to use must be made carefully. It is not necessarily true that your application will perform better with a MultiThreadedExecutor than with a SingleThreadedExecutor just because it can better utilize parallel execution on a multi-core CPU. In the current implementation, a SingleThreadedExecutor is more efficient in terms of CPU usage than a MultiThreadedExecutor. Additionally, it is considered an antipattern to use a MultiThreadedExecutor to run callbacks from a node that should be executed exclusively (we will explore later why this might not always be desired). Lastly, in a `MultiThreadedExecutor`, we may encounter starvation issues for one of the callbacks if the execution time and frequency of the timer and subscriber callbacks are high and their queues are always full. When this happens, a `MultiThreadedExecutor` may only have time to handle one callback for a topic, leaving the other unprocessed.

Let's look at an example that illustrates this issue. To trace its execution, we will use a library called YAETS[2], which employs an asynchronous tracing system to ensure that data collection does not interfere with execution. It also provides with tools to create execution graphs. The `TRACE_EVENT` macro allows us to incrementally log the start and end times of the function in which it is used.

We will run two nodes, Consumer and Producer, within the same Executor:

- **Producer**: This node will publish a message every millisecond to two topics: `topic_1` and `topic_2`.

```cpp
br2_deep_ros/src/executors.cpp

class ProducerNode : public rclcpp::Node
{
public:
  ProducerNode() : Node("producer_node")
  {
    pub_1_ = create_publisher<std_msgs::msg::Int32>("topic_1", 100);
    pub_2_ = create_publisher<std_msgs::msg::Int32>("topic_2", 100);
    timer_ = create_wall_timer(
      1ms, std::bind(&ProducerNode::timer_callback, this));
  }

  void timer_callback()
  {
    message_.data += 1;
    pub_1_->publish(message_);
    message_.data += 1;
    pub_2_->publish(message_);
  }

private:
  rclcpp::Publisher<std_msgs::msg::Int32>::SharedPtr pub_1_, pub_2_;
  rclcpp::TimerBase::SharedPtr timer_;
  std_msgs::msg::Int32 message_;
};
```

- **Consumer**: This node is the one we are truly interested in observing. It has two subscribers, one for `topic_1` and one for `topic_2`, and each of them simulates taking half a millisecond to process using the `WasteTime` method. Additionally, it has a timer that triggers a callback every 10 milliseconds, with the callback taking 5 milliseconds to complete.

---

[2]https://github.com/fmrico/yaets

br2_deep_ros/src/executors.cpp

```cpp
class ConsumerNode : public rclcpp::Node
{
public:
  ConsumerNode() : Node("consumer_node")
  {
    sub_2_ = create_subscription<std_msgs::msg::Int32>(
      "topic_2", 100, std::bind(&ConsumerNode::cb_2, this, _1));
    sub_1_ = create_subscription<std_msgs::msg::Int32>(
      "topic_1", 100, std::bind(&ConsumerNode::cb_1, this, _1));

    timer_ = create_wall_timer(
      10ms, std::bind(&ConsumerNode::timer_cb, this));
  }

  void cb_1(const std_msgs::msg::Int32::SharedPtr msg)
  {
    TRACE_EVENT(session);

    waste_time(500us);
  }

  void cb_2(const std_msgs::msg::Int32::SharedPtr msg)
  {
    TRACE_EVENT(session);

    waste_time(500us);
  }

  void timer_cb()
  {
    TRACE_EVENT(session);

    waste_time(5ms);
  }

  void waste_time(const rclcpp::Duration & duration)
  {
    auto start = now();
    while (now() - start < duration);
  }

private:
  rclcpp::Subscription<std_msgs::msg::Int32>::SharedPtr sub_1_;
  rclcpp::Subscription<std_msgs::msg::Int32>::SharedPtr sub_2_;
  rclcpp::TimerBase::SharedPtr timer_;
};
```

The `main()` function simply creates an executor, adds the nodes to it, and calls spin:

br2_deep_ros/src/executors.cpp

```cpp
int main(int argc, char * argv[])
{
  rclcpp::init(argc, argv);

  auto node_pub = std::make_shared<ProducerNode>();
  auto node_sub1 = std::make_shared<ConsumerNode>();

  rclcpp::executors::SingleThreadedExecutor executor;

  executor.add_node(node_pub);
  executor.add_node(node_sub1);

  executor.spin();

  rclcpp::shutdown();
  return 0;
}
```

Considering the publication frequencies of the producer and the execution times of each callback in the Consumer, we can observe that the Consumer always has work to do, as its message queues are constantly full. Figure 7.6 shows the execution of the callbacks in the Consumer node the first millisconds of its execution. The system where this program was executed is not overloaded, so there are no delays in their execution. Using a `SingleThreadedExecutor`, we can confirm everything we have explained in this section:

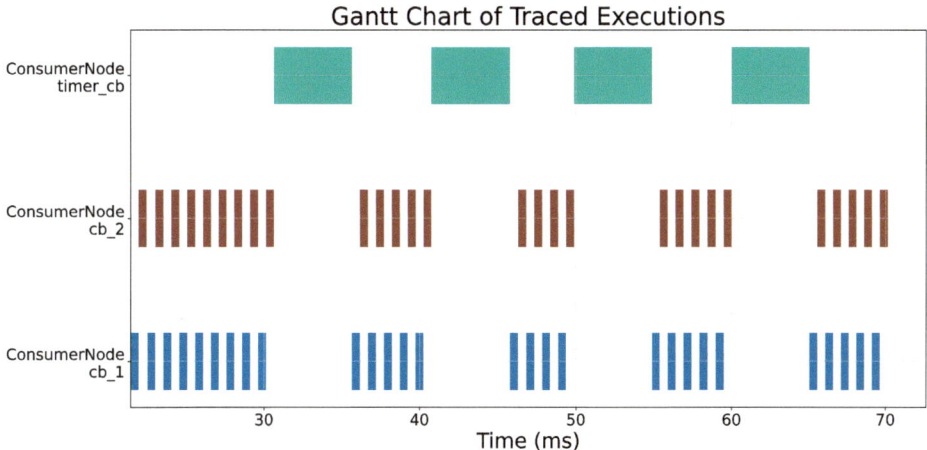

Figure 7.6: Trace of the program execution with a `SingleThreadedExecutor`. It can be observed that no concurrency occurs in the execution of any callback

- The execution of the timer callback takes priority over the subscribers. Even if there are more messages queued, if there is a timer event at the beginning of a processing window, it is the first thing to be executed.

- The first callback to be executed is the one that was created last, which is `cb2`, the callback for `topic_2`.

- No callback is executed concurrently.

Now, let's switch to a `MultiThreadedExecutor` by simply changing the type of executor and configuring it to use 8 threads:

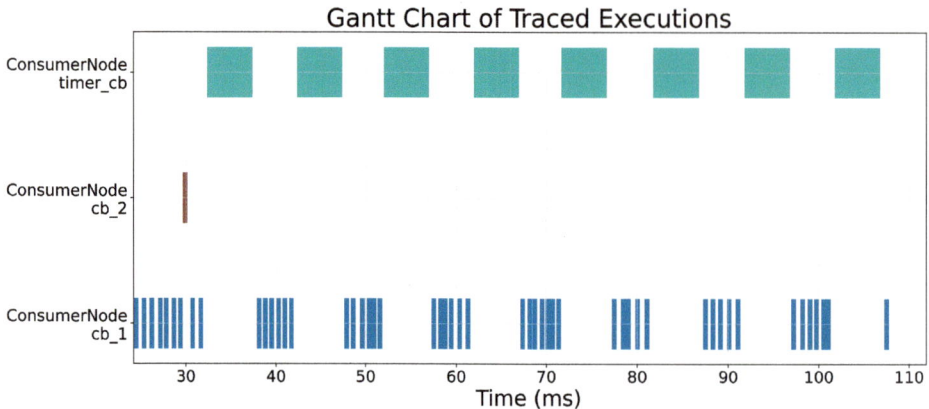

Figure 7.7: Execution trace of the program with a `MultiThreadedExecutor`. It can be observed that no concurrency occurs in the execution of any callback, but the callback for `/topic_2` experiences starvation.

```
br2_deep_ros/src/executors.cpp

    int main(int argc, char * argv[])
    {
      rclcpp::init(argc, argv);

      auto node_pub = std::make_shared<ProducerNode>();
      auto node_sub1 = std::make_shared<ConsumerNode>();

      // rclcpp::executors::SingleThreadedExecutor executor;
      rclcpp::executors::MultiThreadedExecutor executor(rclcpp::ExecutorOptions(), 8);

      executor.add_node(node_pub);
      executor.add_node(node_sub1);

      executor.spin();

      rclcpp::shutdown();
      return 0;
    }
```

Figure 7.7 shows the execution with `MultiThreadedExecutor`.

And we see a discouraging result: once messages begin to accumulate in the queues due to the first execution of the timer callback, the callback for `topic_2` is no longer called. Additionally, it can be observed that, even with multiple threads, within a node, by default, only one callback is executed at a time, meaning there is no concurrency.

It is left as an exercise for the reader to test with the `EventsExecutor`. The result should be similar to the one with the `SingleThreadedExecutor`. To try, simply change the type of executor and reproduce these graphs by running:

```
$ colcon build --symlink-install --packages-up-to br2_deep_ros
$ ros2 run br2_deep_ros executors
$ ros2 run yaets gantt.py ./session1.log --max_traces 60
```

7.1.1 Callback Groups

Up to this point, we have intentionally hidden the concept of *Callback Group* from the reader to avoid confusion while explaining the ROS 2 execution model. However, it is now time to lay all the cards on the table. Executors do not operate at the node level, but at the callback group level. This distinction is usually imperceptible because each node has a default callback group, and all subscribers, timers, and services created within a node are, by default, added to this default callback group. Additionally, this default callback group is configured as *mutually exclusive*, which ensures what we have previously mentioned: that messages and events are processed sequentially, so that no two callbacks from any element within the same callback group are executed simultaneously.

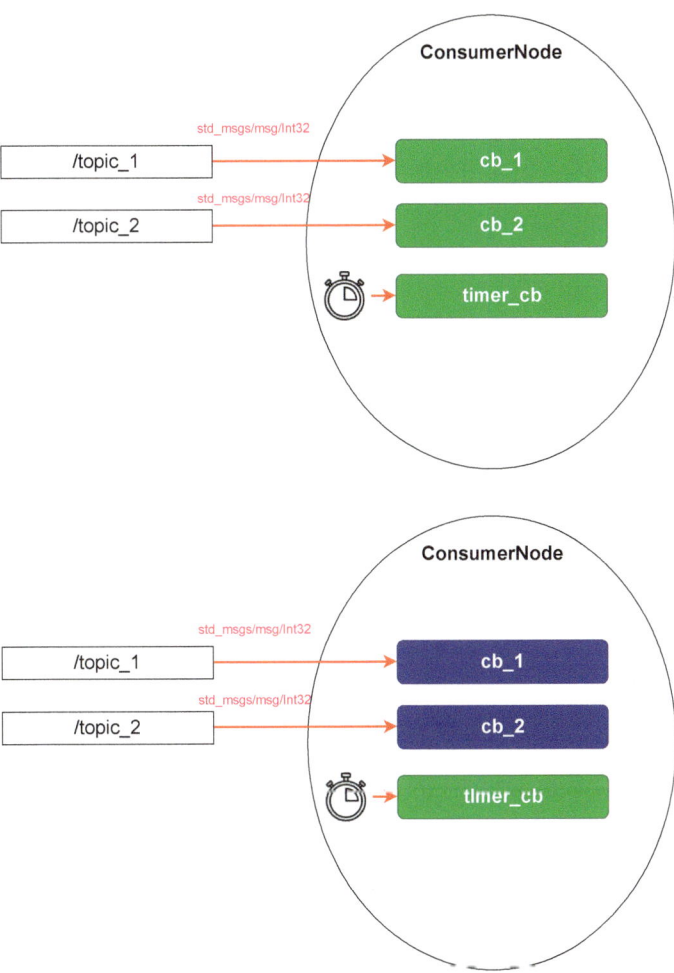

Figure 7.8: Diagram of the `ConsumerNode` with its callbacks, indicating the callback group to which they belong by color. On the top figure, all callbacks are in the default callback group (green). On the bottom figure, the timer callback is in a custom callback group (blue).

In ROS 2, we are allowed to create additional callback groups within a node, selecting which timers, subscribers, or services belong to each callback group, whether it is the default one or any explicitly created ones. We will illustrate this explanation by continuing with the example from the previous section. The `ConsumerNode` in the previous section followed the diagram on the left in Figure 7.8. All callbacks were in the same callback group, which was the default callback group of the node. Now, we will explicitly create an additional callback group within the node, assigning the timer callback to this newly created callback group instead of the default one. To do this, we will use an additional parameter in `create_subscription`, which comes after the ones we typically use. This parameter allows us to configure various subscription options, such as specifying the callback group, enabling intra-process memory, or ignoring local publications, among others. This parameter has a default value, and up until now, we haven't used it.

```cpp
br2_deep_ros/src/executors_cbg.cpp

ConsumerNode() : Node("consumer_node")
{
    custom_cb_ = create_callback_group(rclcpp::CallbackGroupType::MutuallyExclusive);

    rclcpp::SubscriptionOptions options;
    options.callback_group = custom_cb_;

    sub_2_ = create_subscription<std_msgs::msg::Int32>(
      "topic_2", 100, std::bind(&ConsumerNode::cb_2, this, _1), options);
    sub_1_ = create_subscription<std_msgs::msg::Int32>(
      "topic_1", 100, std::bind(&ConsumerNode::cb_1, this, _1), options);

    timer_ = create_wall_timer(10ms, std::bind(&ConsumerNode::timer_cb, this));
}

...

private:
    rclcpp::CallbackGroup::SharedPtr custom_cb_;
```

In the `main()` function, we don't need to do anything differently. When adding a node to an executor, all the callback groups from that node are added to the executor, which in this case will be a `MultiThreadedExecutor`. If we run this program and visualize the execution traces (Figure 7.9), we will see that the callbacks for the topics are still not executed concurrently (with some starvation in `cb_2`), but the timer callback is executed concurrently with these two because it belongs to a different callback group.

Furthermore, we can configure some of these additional callback groups as *reentrant*, as opposed to *mutually exclusive*, meaning that nothing prevents a `MultiThreadedExecutor` from calling the same callback with different data simultaneously. This must be done with great care and often requires the use of concurrency control mechanisms such as mutexes and locks.

In the following code, we make use of this type of reentrant callback groups. We will add all the callbacks to this callback group. If we do nothing else, all callbacks could execute concurrently, but we will use a mutex to prevent the timer callback and the `/topic_1` callback from running at the same time.

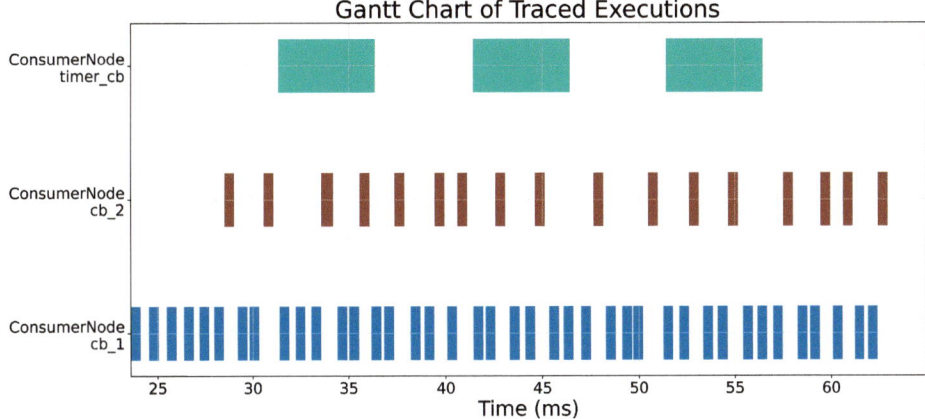

Figure 7.9: Execution flow of the `ConsumerNode`, using two callback groups: the default group for the subscriptions and an additional group for the timer callback. It can be observed that the timer callback runs concurrently with the other two callbacks, which execute without concurrency between them.

```
br2_deep_ros/src/executors_cbg_reentrant.cpp

    ConsumerNode() : Node("consumer_node")
    {
      custom_cb_ = create_callback_group(rclcpp::CallbackGroupType::Reentrant);

      rclcpp::SubscriptionOptions options;
      options.callback_group = custom_cb_;

      sub_2_ = create_subscription<std_msgs::msg::Int32>(
        "topic_2", 100, std::bind(&ConsumerNode::cb_2, this, _1), options);
      sub_1_ = create_subscription<std_msgs::msg::Int32>(
        "topic_1", 100, std::bind(&ConsumerNode::cb_1, this, _1), options);

      timer_ = create_wall_timer(
        10ms, std::bind(&ConsumerNode::timer_cb, this), custom_cb_);
    }

    void cb_1(const std_msgs::msg::Int32::SharedPtr msg)
    {
      std::unique_lock<std::mutex> lock(mutex_);
      TRACE_EVENT(session);

      waste_time(500us);
    }

    void cb_2(const std_msgs::msg::Int32::SharedPtr msg)
    {
      TRACE_EVENT(session);

      waste_time(500us);
    }

    void timer_cb()
    {
      std::unique_lock<std::mutex> lock(mutex_);
      TRACE_EVENT(session);

      waste_time(5ms);
    }
    ...

  private:
    ...
    rclcpp::CallbackGroup::SharedPtr custom_cb_;
    std::mutex mutex_;
```

- When creating the callback group, we set it to be `Reentrant` instead of `MutuallyExclusive`.

- Just as we introduced the additional parameter for `create_subscription`, `create_wall_timer` also has an extra parameter beyond those we typically use, which allows us to directly specify the callback group for the timer.

- We will use a mutex to prevent the timer callback and the `/topic_1` callback from running simultaneously, so we declare a mutex and acquire the lock using `std::unique_lock<std::mutex>`. When this object is created, it acquires the lock if it hasn't been taken by another thread, otherwise it waits.

The execution diagram of this example is shown in Figure 7.10.

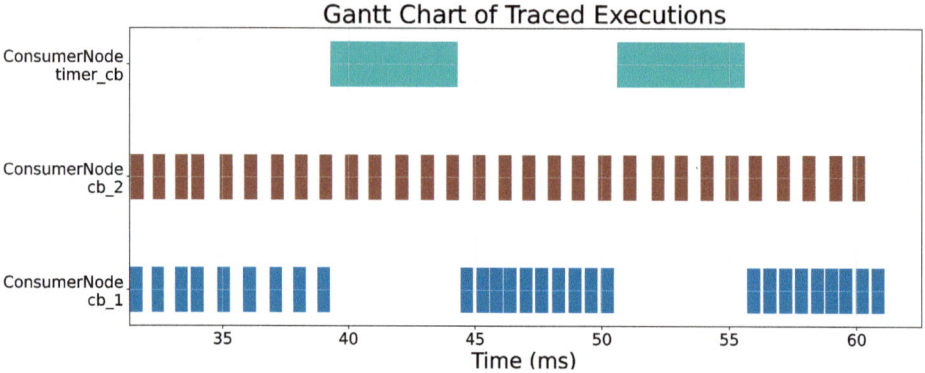

Figure 7.10: Trace of the execution of the `ConsumerNode` with a callback group configured as reentrant, and using a mutex.

7.2 REAL-TIME IN ROS 2

In this section, we will dive into the fascinating world of Real-Time, something that a robotics software engineer often does not think about at first. It is usually the specific needs of your application or your experience with undesirable effects that make you realize its importance and that you need to, at the very least, understand the basic techniques of real-time computing to ensure your applications run correctly. In this chapter, we have learned about executors and callback groups, which we will soon see are essential ingredients for making your application run in real-time.

Join me on this journey, starting with the definitions of what we mean by real-time and ending with how we make our ROS applications run in real-time.

7.2.1 About Real-Time

The term *Real-Time* is an ambiguous one in computing, as it has been used in many domains and in various ways. A web application developer for food delivery might tell you that their app allows tracking of orders in real-time. A security application developer might claim that their amazing mobile app lets you watch your cats in real-time while you're on vacation in southern Spain. A simulation programmer might tell you that a particular clock runs in real-time. Expert hackers might say that certain operating systems are real-time, while other programmers will assert that their applications run in real-time. Which of these statements is correct? In fact, all of them are correct within the context of each application, so we must clarify which type of real-time we are going to discuss in this book. The definition of real-time that we will use in this book is:

*Real-time programming refers to the practice of designing and implementing software systems that can respond and produce output within **strict time constraints**.*

When classifying real-time systems, two criteria are considered:

1. A system must respond within a specified time frame, where the maximum latency is limited.

2. The consequences of failing to meet the established deadlines. These consequences can range from something as mild as a brief interruption in playing your favorite song to something as serious as hitting your neighbor with a vehicle, or as catastrophic as a plane with 200 people crashing.

In Figure 7.11, we see a graph where different applications or use cases are plotted on severity/latency coordinates. The control loop of actuators, such as those in vehicles or planes, typically requires very high frequencies, close to 1000 Hz. If this control loop becomes unstable or fails to meet the required frequency, it can create unsafe conditions, potentially leading to catastrophic consequences, including endangering the lives of the passengers. The braking system of a car also carries very severe consequences if more than 10 milliseconds pass between pressing the brake pedal and the brake being applied to the wheels. Robots must react to sensor information within short time deadlines if they are to avoid obstacles when moving at a certain speed. Music production systems also require low-latency components, although no one will die if these deadlines are not met, even though the financial costs of such an error could be significant. On the other hand, in a video conference, if latency exceeds half a second, communication becomes difficult, but no lives are at risk, and there are no major financial costs involved.

In this same figure, a boundary encloses the applications or use cases that can be executed on general-purpose operating systems, such as Linux, for example. Anything outside this boundary requires the use of real-time operating systems, such as QNX, NuttX, VXWorks, or systems with hypervisors like Xenomai. The humanoid robot Nao, for instance, had a general-purpose computer in its head, but the functions that required real-time execution are handled by a microcontroller located in the robot's chest, which ran in real-time.

The reason that general-purpose operating systems are not well-suited for real-time applications lies in their process schedulers, which allow processes controlling

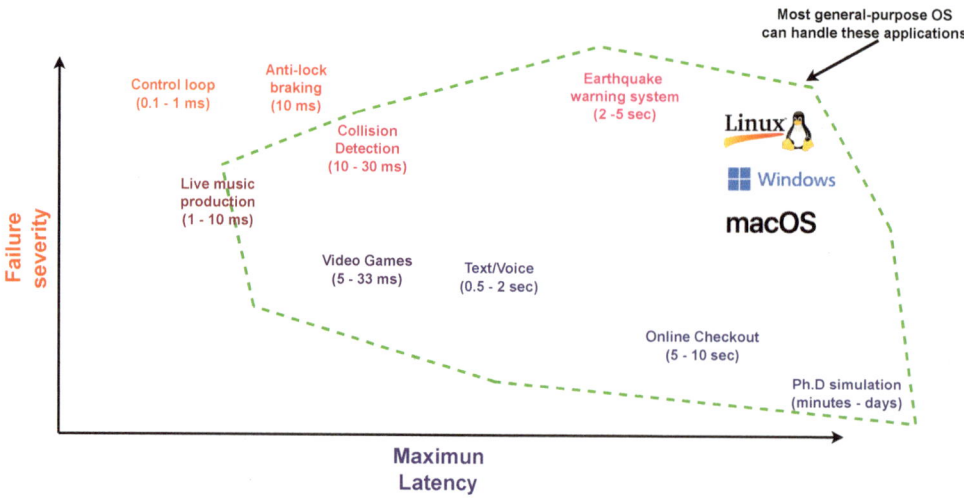

Figure 7.11: Example of Real-Time use cases in a Severity/Latency graph. The dashed green line delineates the use cases that can be handled by general-purpose operating systems.

real-time applications to be preempted from the CPU for an indefinite amount of time while attending to other processes that may be of relative importance, such as handling keyboard input, refreshing a screen widget, or receiving an email notification. Even if we explicitly increase the priority of these processes, kernel interrupts can still unjustifiably delay a real-time process. One solution to this issue is to use the PREEMPT_RT patch for the Linux kernel, which is included in kernels released after September 2024. Many companies (such as SpaceX for flight software, robotic arms, and autonomous vehicle companies), although some do not publicly admit it, find this patched Linux kernel sufficient for their real-time needs.

7.2.2 System Latencies

Latency is the primary metric when evaluating a real-time system. It is defined as the time that passes between the occurrence of an event and the system's reaction to it. For example, if we consider driving a car as a system that includes both the driver and the vehicle, the time that elapses from the moment we see an obstacle on the road to when the car's brakes begin to engage would be the latency in response to the event of perceiving the obstacle. In the case of a robot navigating to avoid obstacles, latency refers to the time between perceiving an obstacle and sending the velocity commands generated to avoid it. In both cases, latency indicates how quickly the system reacts, and avoiding a collision depends on keeping that latency from being excessive.

Modern software stacks are very deep, and each layer independently adds its own latency to the total response time. In real-time applications, it is necessary to guarantee a maximum latency, which is challenging because most existing software is designed to minimize average latency, not maximum latency. In any case, it is

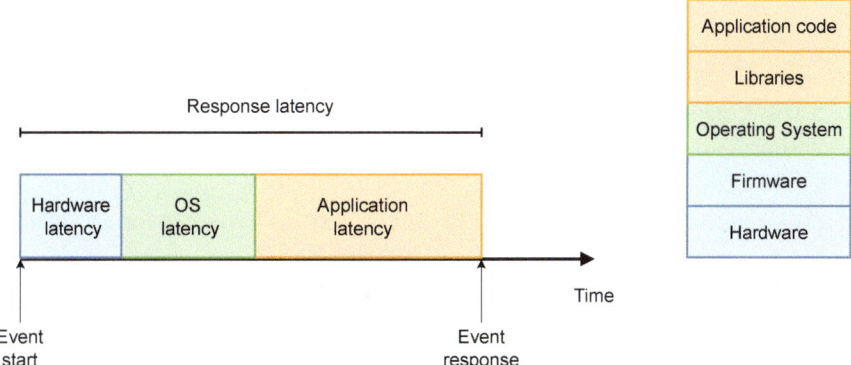

Figure 7.12: Response latency is composed of the different latencies in the various layers of the system.

essential to understand where the latencies come from and to measure the maximum latency that we are handling in the system we are using.

As shown in Figure 7.12, response latency is composed of multiple sources of latency:

- **Application latency** is the sum of the application's own latency and the latency added by the libraries it uses. The quality of the implementation of the libraries we use and how we program our applications define this latency.

- **Operating system latency** primarily stems from how the operating system schedules processes.

- **Hardware latency** depends on both physical variables and the programming of the system components' firmware.

Modern hardware is very complex, and the operating system does not fully control it. System Management Interrupts (SMI) and simultaneous multithreading make it difficult to manage hardware behavior. Additionally, factors such as power, which dynamically scales hardware frequency, or temperature, can affect performance.

Let us perform a small exercise to measure the latency induced by SMIs by running the `hwlatdetect` command. Ideally, to validate a system with a high level of confidence and determine these maximum latencies, this test should run for several days, and it should always be done with the system under stress, for which we will use the `stress-ng` command. You can monitor the system usage by running the `htop` command in another terminal.

```
$ htop # (optional) in terminal 0
$ stress-ng -c $(nproc) -t 300 # In terminal 1
$ sudo hwlatdetect # In terminal 2
```

In my case, I am writing this book on a Lenovo ThinkPad P14s Gen 3 with a 12th Gen Intel Core i7-1280P processor with 20 cores and 48 GiB of RAM, and the maximum latency detected is 27 microseconds (µs). If we had found maximum latencies close to a millisecond due to periodic hiccups in our computer, it would have been time to consider purchasing a more suitable machine for real-time applications.

Cyclictest is another interesting command. Cyclictest measures the real-time latency of the operating system by evaluating how long it takes for a thread to wake up after a scheduled timer. It works by generating periodic timer events and recording the difference between the expected and actual time, allowing for the identification of delays in process scheduling.

These are system tools, but next we will present an example that does something similar to Cyclictest, which will help us better understand the concepts in this chapter. Let us look at this simple program:

```cpp
br2_deep_ros/src/wakeup.cpp

#include "yaets/tracing.hpp"

int main(int argc, char * argv[])
{
  yaets::TraceSession session("wakeup.log");

  while(true) {
    {
      TRACE_EVENT(session);
    }
    std::this_thread::sleep_for(std::chrono::milliseconds(20));
  }

  return 0;
}
```

In this program, we are not as concerned with showing the sequence of executions but rather the time between executions of each loop cycle. If this program had no underlying latency, it would always run every 20 milliseconds. YAETS will help us measure whether this is the case. Remember that TRACE_EVENT stores the timestamp asynchronously (to avoid affecting the program's behavior) when a block starts and ends (simplifying). Let's execute:

```
$ stress-ng -c $(nproc) -t 300 # In terminal 0
$ ros2 run br2_deep_ros wakeup # In terminal 1
```

To see the time between TRACE_EVENT calls, we will use the following command:

```
$ ros2 run yaets elapsed_time_histogram.py wakeup.log --function main --bins 40
```

After running for about 1 minute, stop al commands via Ctrl-C. Figure 7.13 shows the histogram of the loop durations. We can see that the vast majority of iterations last the expected 20 milliseconds. However, there are some iterations that, due to scheduler-induced latency, last up to 27 milliseconds, which is significant.

Any of these iterations that occur later than expected could be the one that prevents us from reacting in time, potentially causing a controller to become unstable, a robot to collide with an obstacle, or a plane to crash. And in this case, where there are no additional libraries or code involved, all of these delays occur because the operating system preferred to allocate CPU time to a different process instead of the

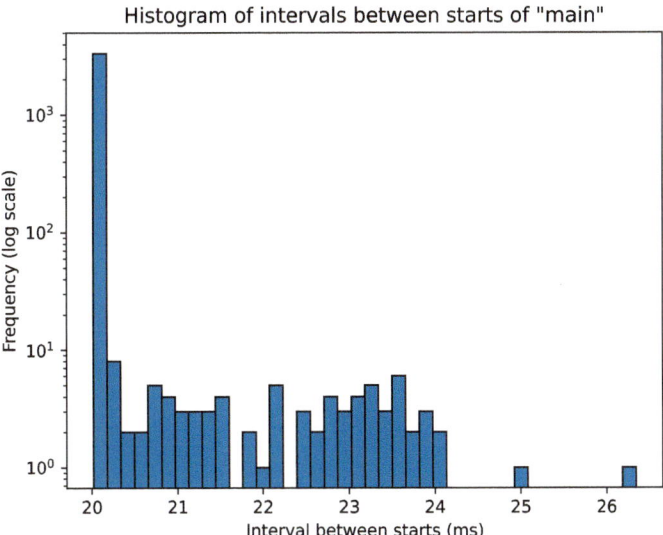

Figure 7.13: The histogram shows the time between executions of each loop cycle. The Y-axis, in logarithmic scale, represents the number of times a cycle lasted the duration shown on the X-axis.

one that really needed to react in time. There is a very straightforward way to fix this problem in this example, but first we need to discuss how the Linux kernel scheduler works and what mechanisms it offers to indicate that your process is more important than most others and should be given priority.

7.2.3 Dealing with the Operating System Scheduler

Process scheduling is a very complex problem, classified as NP-complete. The Linux kernel includes several schedulers, with the default one being the Completely Fair Scheduler (CFS), whose policy name is SCHED_OTHER. The CFS algorithm optimizes the average response time, which is suitable in the vast majority of cases for general-purpose computer usage. This scheduler is the one that managed the execution of the wakeup program shown earlier. However, this scheduler is not suitable for real-time applications. There are two alternatives that are suitable for real-time systems:

- A **FIFO Scheduler** (SCHED_FIFO) assigns the CPU to processes in the order they arrive (First In, First Out), without preemption. Once a process obtains the CPU, it continues to run until it finishes or blocks, and only then is the CPU assigned to the next process in the queue. It is ideal for critical applications where predictable execution is required.

- A **EDF Scheduler** (SCHED_DEADLINE) assigns the CPU to processes based on their deadlines (Earliest Deadline First). Each task has an execution time, a period, and a deadline, and the scheduler ensures that tasks are executed before their deadline, prioritizing those with the nearest deadline. It is ideal for critical applications where meeting timing requirements is essential.

Using an EDF scheduler is somewhat complex because you have to specify the deadline, but using a FIFO scheduler is quite simple. You only need to specify a priority. But how are these priorities set?

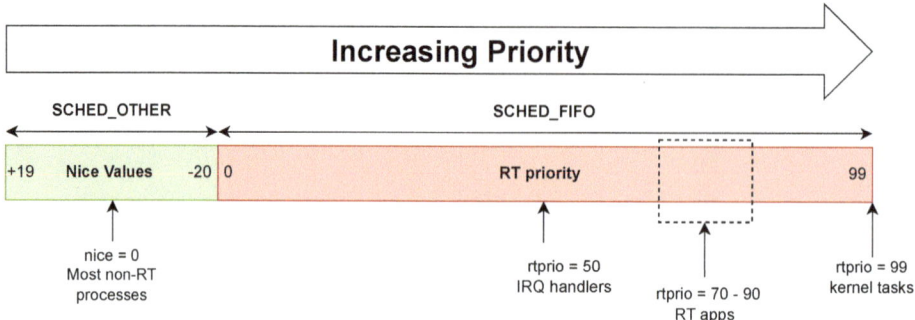

Figure 7.14: How to interpret the `nice` and `rtprio` values for the `SCHED_OTHER` and `SCHED_FIFO` schedulers, from left (lowest priority) to right (highest priority).

When using the `SCHED_OTHER` scheduler, it is possible to give a process a slight advantage over another by setting a `nice` value, which is a number between $+19$ (lower priority) and -20 (higher priority), as shown in Figure 7.14. All processes scheduled with `SCHED_FIFO` are prioritized over those scheduled with `SCHED_OTHER` (even if their nice value is -20). When using `SCHED_FIFO`, we set the `rtprio value`, ranging from 0 (lower priority) to 99 (higher priority). A value of 80 is typically suitable for a real-time application like the ones we will use in this chapter.

Let's see how we can designate our wakeup program as a real-time application. You need to give permissions to your user to increase a thread's priority. In my case, since my user is *fmrico*, I will create a file in `/etc/security/limits.d` called `20-fmrico-rtprio.conf` (the file name is not very important, and the number before the name is simply to prioritize certain files in the directory over others in case of conflict) with the following content. This allows the user *fmrico* to raise the priority of a process up to the value 98.

```
/etc/security/limits.d/20-fmrico-rtprio.conf

    fmrico - rtprio 98
```

If you have never done this setup before, do this change in `20-fmrico-rtprio.conf`. While you have not done any action to check or set this, read again previous sentence.

I trust that the previous paragraph has not caused an infinite loop for the reader, and that we can continue. Now, let us see how to modify `wakeup.cpp` to increase its priority. The function `sched_setscheduler` allows us to set the scheduler and parameters (such as the `rtprio` priority) for the thread in which it is called.

```
br2_deep_ros/src/wakeup.cpp
```

```cpp
#include <stdexcept>
#include <cstring>

int main(int argc, char * argv[])
{
  yaets::TraceSession session("wakeup.log");

  sched_param sch;
  sch.sched_priority = 80;
  if (sched_setscheduler(0, SCHED_FIFO, &sch) == -1) {
    throw std::runtime_error{
      std::string("failed to set scheduler: ") +
      std::strerror(errno)};
  }

  while(true) {
    {
      TRACE_EVENT(session);
    }
    std::this_thread::sleep_for(std::chrono::milliseconds(20));
  }

  return 0;
}
```

Let us check if our program has improved by generating a new histogram of the execution latencies:

```
$ stress-ng -c $(nproc) -t 300 # In terminal 0
$ ros2 run br2_deep_ros wakeup # In terminal 1 for 1 minute

$ ros2 run yaets elapsed_time_histogram.py wakeup.log --function main --bins 40
```

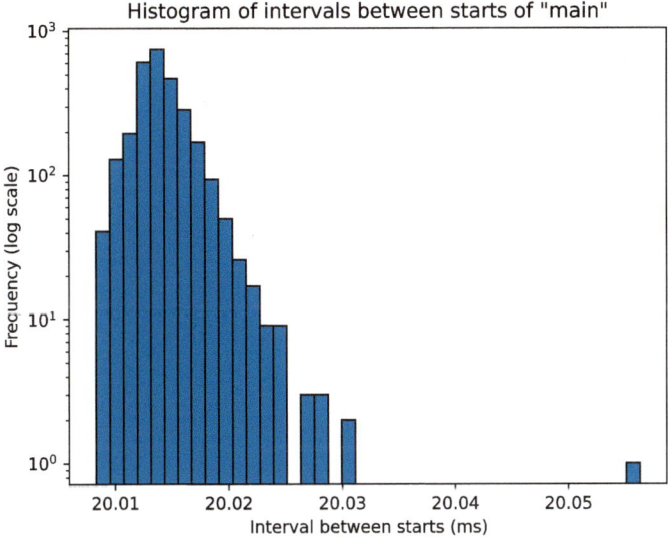

Figure 7.15: The histogram shows the time between executions of each loop cycle using a SCHED_FIFO scheduler and a rtprio = 80.

Honestly, I would love to be beside the reader at this moment to see their *WTF!* expression upon realizing that our program is now much more deterministic, by a large margin. What used to have a range of up to 27 milliseconds in Figure 7.15 now never exceeds 20.06 milliseconds. What happened? Now, the only things separating you from the desired 20 milliseconds are hardware interrupts and kernel processes, no poor, middle-class processes.

7.2.4 Considerations for Generic Real-Time Programming without ROS

In these sections, we are indicating that in order to have a real-time application, it is necessary to increase the priority of the tasks we have identified as real-time tasks. These may be all the tasks in our application or perhaps just a subset of them. These techniques, along with those we will explore in later sections—directly using ROS 2 mechanisms—will ensure that our code is predictable and meets its tasks within strict time constraints. However, we must pause for a moment to present other factors that can cause delays in the execution of our code due to various issues. These are general programming considerations, and although we will not address them beyond this section, it is important to present them as they are critical when developing real-time applications in any domain. We will not present complex examples, as they would require lengthy and detailed explanations, and, after all, this is a book on ROS 2 programming, not one specifically about real-time application development.

7.2.4.1 *Page Faults*

In real-time applications, **page faults** can generate unwanted latencies because they involve loading data from secondary memory (disk) to primary memory (RAM). This can result in the transfer of large amounts of data between memory and disk, and although solid-state drives have made significant advances in recent years, they can still bring our application to apply the handbrake.

For example, a simple `return` statement in a function could cause a page fault. Every time a function is called, space is reserved on the stack to store local variables and the return address. When the function ends and a return is executed, the CPU accesses the stack to obtain the return address (where the program execution should continue). If the memory page containing the stack was swapped to disk due to a lack of recent use, accessing that address triggers a page fault. Additionally, if the function has local variables that have not been accessed for some time, and those variables are stored on a page that was evicted from RAM, an implicit access to these variables during the return (such as clearing or updating the stack) can also cause a page fault.

In a real-time system, these faults are critical because the system must guarantee predictable response times. To prevent a return from causing page faults, techniques such as locking pages in memory (page locking) can be used, through system calls like `mlock()` or `mlockall()`, which prevent relevant pages from being swapped to disk. By calling `mlockall(MCL_CURRENT | MCL_FUTURE)` at the start of your application, demand paging and swapping are turned off for that application.

7.2.4.2 Dynamic Memory

The next consideration is the **use of dynamic memory**. Any engineer who has worked with real-time systems will tell you that using dynamic memory is strictly prohibited in these applications. We are very used to creating temporary images or adding elements to a list without considering that memory allocation is usually not $O(1)$, and can trigger page faults even if `mlockall(MCL_CURRENT | MCL_FUTURE)` is called.

A simple `my_vector.push_back(data)` executed frequently in a real-time task can be unacceptable in terms of predictability and latency. One strategy to avoid this is to use a memory allocator with $O(1)$ complexity, such as the TLSF allocator, which has an example of its usage in one of the tutorials in the ROS 2 documentation[3]. Another strategy is to pre-allocate all the memory you anticipate needing at the start of the application, and use no memory beyond that. For example, in the case of `std::vector`, you can use its `my_vector.reserve(N)` method, which allocates memory for future `push_back` operations. For large data objects, having a memory pool pre-allocated from the beginning of the application, and not using any other memory, can be an effective approach.

7.2.4.3 Priority Inversion in Mutexes

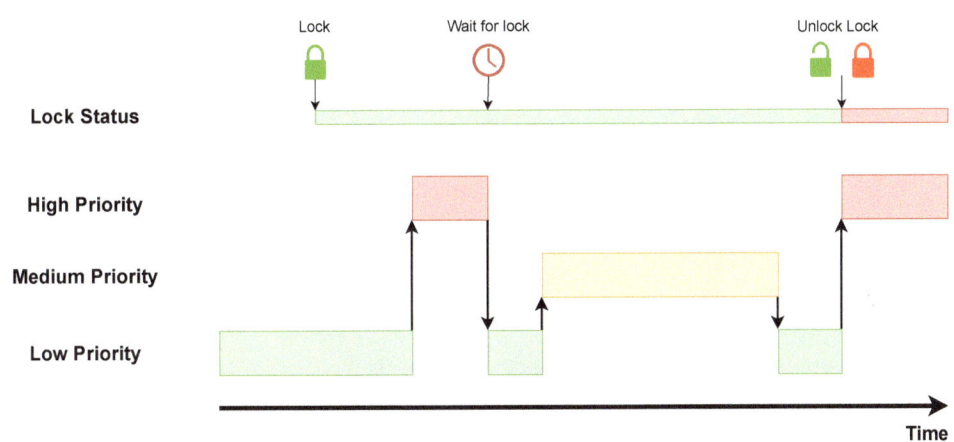

Figure 7.16: The problem of priority inversion. Execution of three processes with different priorities. The lock state is shown at the top, with the color of the process holding it at each moment.

Real-time engineers will also tell you to avoid using `std::mutex` because it can lead to *priority inversion*. Let us consider the following scenario to illustrate this explanation, shown in Figure 7.16: We have a system with three processes—one high-priority, one low-priority, and another medium-priority. At a given moment, the low-

---

[3]https://docs.ros.org/en/rolling/Tutorials/Advanced/Allocator-Template-Tutorial.html#the-tlsf-allocator

priority process is running and acquires a lock it shares with the high-priority process for synchronization. Shortly after, the high-priority process needs to run and preempts the low-priority process. After a brief period, it requires the lock that the low-priority process had acquired, so it suspends its execution until the lock is released. When we return to the low-priority process, it is preempted by the medium-priority process, which is now due to run, delaying the low-priority process from releasing the lock as quickly as possible, preventing the high-priority process from progressing. Priority inversion occurs because a process that does not hold any lock is now executing freely, delaying the execution of higher-priority processes.

One possible solution is to use priority inheritance mutexes, which is a mutex that causes the low-priority process to execute at the priority level of the highest-priority process waiting for the mutex held by the low-priority process. This prevents the medium-priority process from delaying the release of the lock, as shown in Figure 7.17.

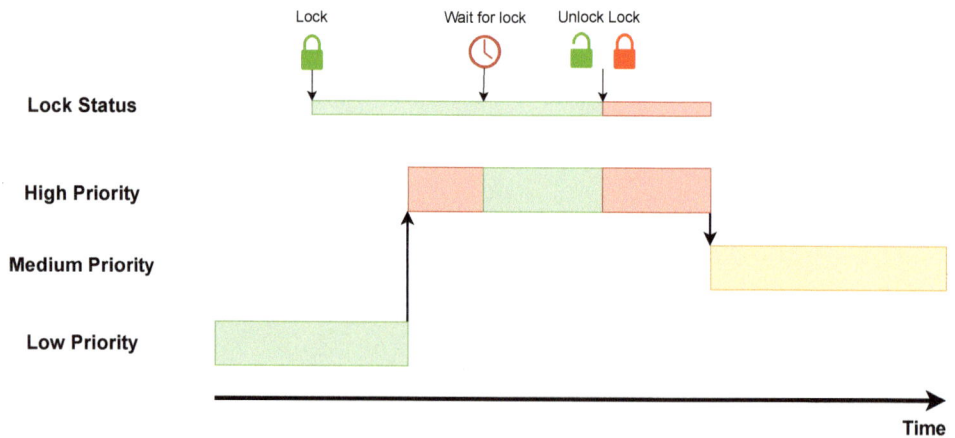

Figure 7.17: Possible solution of priority Inversion using a priority inheritance mutex.

The C++ standard library does not support priority inheritance mutexes, but implementing them is quite straightforward. Here there is a simple one, taken simplified from an implementation in cactus-rt[4]:

```cpp
class mutex {
  pthread_mutex_t m_;

public:
  using native_handle_type = pthread_mutex_t*;

  mutex() {
    pthread_mutexattr_t attr;

    int res = pthread_mutexattr_init(&attr);
    res = pthread_mutexattr_setprotocol(&attr, PTHREAD_PRIO_INHERIT);
    res = pthread_mutex_init(&m_, &attr);
  }
```

---

[4]https://github.com/cactusdynamics/cactus-rt/blob/master/include/cactus_rt/mutex.h

```
~mutex() {
  pthread_mutex_destroy(&m_);
}

mutex(const mutex&) = delete;
mutex& operator=(const mutex&) = delete;

void lock() {
  auto res = pthread_mutex_lock(&m_);
}

void unlock() noexcept {
  pthread_mutex_unlock(&m_);
}

bool try_lock() noexcept {
  return pthread_mutex_trylock(&m_) == 0;
}

native_handle_type native_handle() noexcept {
  return &m_;
}
};
```

Another solution is to avoid using mutexes altogether and apply *lockless programming* techniques, which is an advanced technique to safely share data between cores without the use of locks. This approach requires a redesign of the application and the use of atomic variables when the data to be shared is small, or structures like a ring buffer when the data is larger.

7.2.4.4 Use of I/O

Using input/output, whether to the screen or to files, in a real-time thread is a bad idea, no matter how you look at it. If it involves file input/output, you have to account for the latencies of slower memory and the uncertainties of a file system that you do not control. Screen input/output also introduces latencies and uncertainties that are not acceptable. Despite this, logging is necessary in applications due to its utility or legal requirements. There are asynchronous logging techniques (as used by YAETS) that allow decoupling the real-time thread that generates log data from the thread that actually performs the input/output to the screen or disk.

7.2.5 Real-Time Strategies in ROS 2

Incredibly, as of today, ROS 2 does not have explicit real-time mechanisms, leaving this entirely in the hands of the application programmer and the operating system. Although there are efforts to provide executors with more powerful APIs that can support priority specification, the truth is that these are not yet available in the standard executors. Typically, all callbacks are placed in one executor, which means one or more threads with the same priority, under the same conditions as the other threads/processes in the system. Figure 7.18 shows several callbacks in the same callback group, executing without considering whether any processing needs to be done in real-time.

The strategies we will present below are essentially based on ensuring that callbacks we consider real-time are executed on higher-priority threads, while less important callbacks run on threads with standard priority. Figure 7.19 shows that a

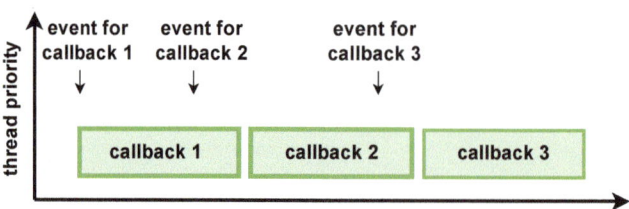

Figure 7.18: All the callbacks execute in a thread with the same priority.

callback, which we consider to have real-time requirements, should be executed in a callback group with a thread with higher priority than the other callbacks without such requirements. This ensures that when the event it needs to process occurs, the lower-priority callback can even be preempted to meet the real-time requirements.

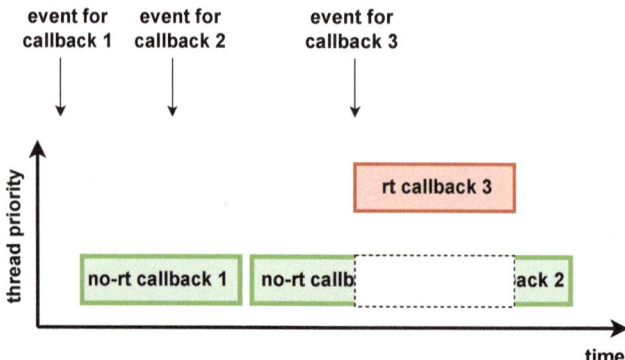

Figure 7.19: Real-time callbacks execute at higher priority, preempting low-priority callbacks.

7.2.5.1 Strategy 1: Nodes with Different Priority

The first strategy is the simplest, where we identify which nodes in our application have higher priority than others, as their callback and timer calls cannot be delayed or interrupted by other system processes. In this case, the strategy involves using two executors: one for the critical nodes that run on a high-priority thread, and another for the other nodes that run on normal-priority threads.

Let us look at an example, illustrated in Figure 7.20, where there are three nodes: a data-producing node and two consumer nodes. One of the consumer nodes must perform critical work with the received data. The other node simply logs the information and has no real-time requirements. We consider both the producer node and the logger node as non-critical and place them in an executor that runs on a standard-priority thread. For the consumer node, we will use an executor that runs on a real-time thread.

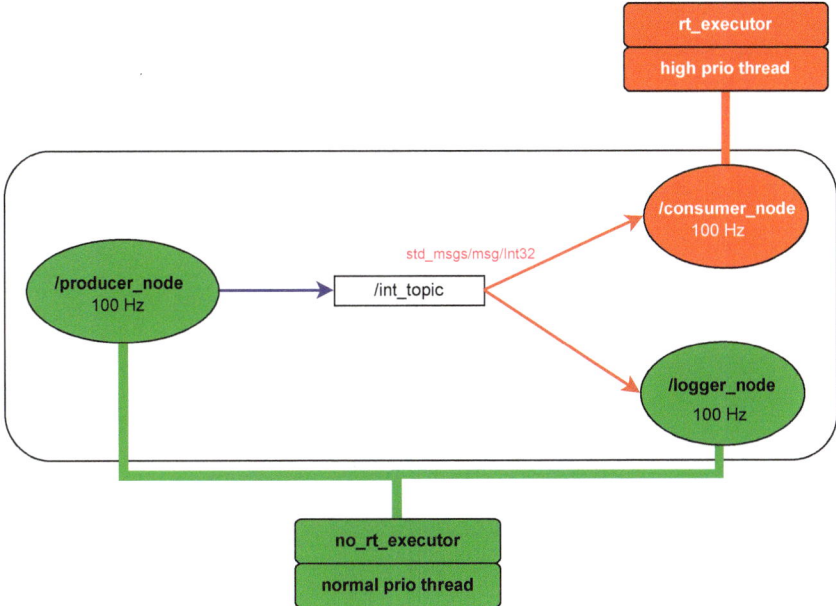

Figure 7.20: Application graph with the Real Time node and executor in red, and the standard nodes and executor in green.

- The `ProducerNode` is a node with a timer that publishes messages at 100 Hz with an incremental counter.

```cpp
br2_deep_ros/src/strategy_1.cpp

class ProducerNode : public rclcpp::Node
{
public:
  ProducerNode() : Node("producer_node")
  {
    pub_ = create_publisher<std_msgs::msg::Int32>("int_topic", 100);
    timer_ = create_wall_timer(10ms, std::bind(
      &ProducerNode::timer_callback, this));
  }

  void timer_callback()
  {
    TRACE_EVENT(session);
    waste_time(shared_from_this(), 200us);

    message_.data += 1;
    pub_->publish(message_);
  }

private:
  rclcpp::Publisher<std_msgs::msg::Int32>::SharedPtr pub_;
  rclcpp::TimerBase::SharedPtr timer_;
  std_msgs::msg::Int32 message_;
};
```

- The `ConsumerNode` subscribes to the topic where the `ProducerNode` publishes, spending some time in the callback that processes it. It also has a timer for a task specific to this node.

```
br2_deep_ros/src/strategy_1.cpp

class ConsumerNode : public rclcpp::Node
{
public:
  ConsumerNode() : Node("consumer_node")
  {
    sub_ = create_subscription<std_msgs::msg::Int32>(
      "int_topic", 100, std::bind(&ConsumerNode::cb, this, _1));
    timer_ = create_wall_timer(10ms, std::bind(
      &ConsumerNode::timer_cb, this));
  }

  void cb(const std_msgs::msg::Int32::SharedPtr msg)
  {
    TRACE_EVENT(session);
    waste_time(shared_from_this(), 500us);
  }

  void timer_cb()
  {
    TRACE_EVENT(session);
    waste_time(shared_from_this(), 2ms);
  }

private:
  rclcpp::Subscription<std_msgs::msg::Int32>::SharedPtr sub_;
  rclcpp::TimerBase::SharedPtr timer_;
};
```

- The `LoggerNode` has exactly the same implementation as the `ConsumerNode`, so we will omit its listing. The reader can check it in the book's repository if desired.

- The key and interesting part of this example is in the `main()` function, where the nodes and executors are created, paired, and execution begins:

  1. Each node is created in the usual way. Two executors are created: one real-time and one non-real-time, and each node is assigned to one of the executors. Only the ConsumerNode is placed in the real-time executor.

```
br2_deep_ros/src/strategy_1.cpp

int main(int argc, char * argv[])
{
  rclcpp::init(argc, argv);

  auto node_producer = std::make_shared<ProducerNode>();
  auto node_consumer = std::make_shared<ConsumerNode>();
  auto node_logger = std::make_shared<LoggerNode>();

  rclcpp::executors::SingleThreadedExecutor no_rt_executor;
  rclcpp::executors::SingleThreadedExecutor rt_executor;

  no_rt_executor.add_node(node_producer);
  no_rt_executor.add_node(node_logger);
  rt_executor.add_node(node_consumer);
```

  2. Next, a thread is created where `spin()` is called on the real-time executor immediately after setting the thread to `SCHED_FIFO` with a priority of 90. This thread is real-time. The main thread continues in `main()`, and we call `spin()` in the usual way.

When the execution of the executors completes, we need to be careful to wait for all threads to finish. Therefore, we join the real-time thread.

```
br2_deep_ros/src/strategy_1.cpp

auto rt_thread = std::thread(
  [&]() {
    sched_param sch;
    sch.sched_priority = 90;

    if (sched_setscheduler(0, SCHED_FIFO, &sch) == -1) {
      throw std::runtime_error{
        std::string("failed to set scheduler:") + std::strerror(errno)};
    }

    rt_executor.spin();
});

no_rt_executor.spin();

rt_thread.join();

rclcpp::shutdown();
return 0;
}
```

The goal of this strategy (Figure 7.21) is:

- To improve the **bounded update rate**. The aim is for the execution of timer callbacks to adhere to the set frequency, avoiding delays or early executions, thus making execution more predictable.

- To reduce the **response time.**It is required that the time between the start and end of a callback's execution does not occasionally increase due to system scheduling issues.

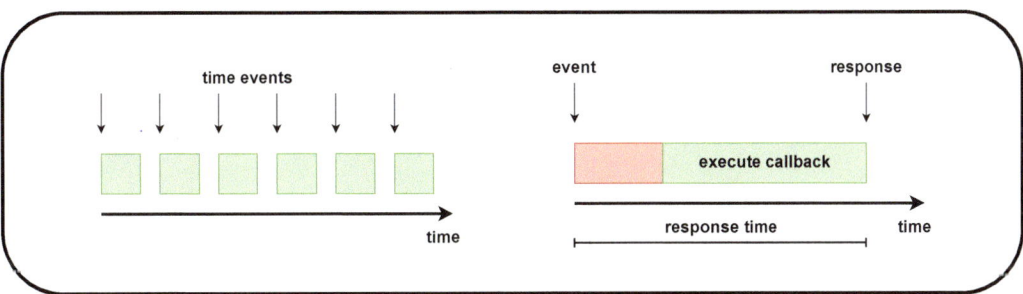

Figure 7.21: Benefits of strategy 1: Improves bounded update rate and reduces response time.

To run and measure the program, the first step is to stress the system. I want to emphasize that this is important in these examples, as in a lightly loaded system, the benefits of real-time techniques may not seem as apparent. The same happens during software development: when we test everything in isolation, everything seems to work fine. The problem arises when we integrate everything, and suddenly the entire structure collapses. We realize that some parts of our application can degrade, such as debug information or certain services, but those parts with strict time constraints cannot degrade. If they do not meet these constraints, the robot's functionality fails completely. It is something we often overlook until we reach deployment on the real robot.

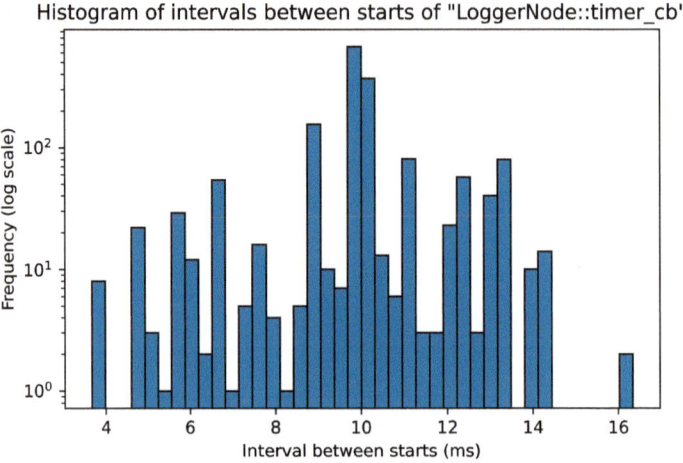

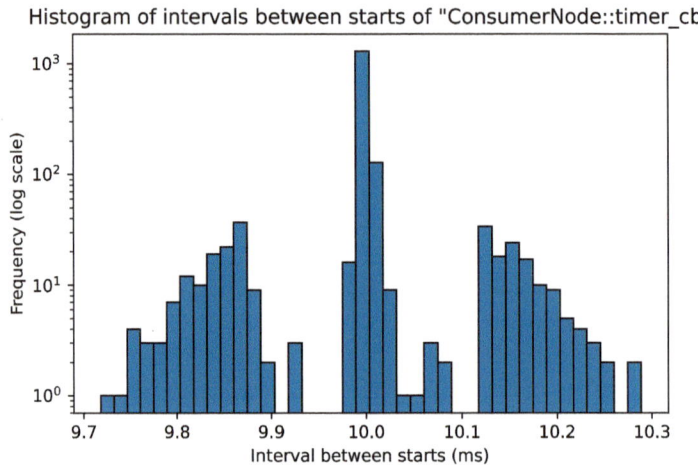

Figure 7.22: Time between executions of the timer callbacks for the `LoggerNode` and the `ConsumerNode`. It is important to observe the scale of the X-axis to see the differences.

To stress the system and run the program, we use the following commands in different terminals:

```
$ stress-ng -c $(nproc) -t 300 # In terminal 1
```

```
$ ros2 run br2_deep_ros strategy_1 # In terminal 2
```

After about 15 seconds, stop both processes with Ctrl-C, and let us analyze the execution. We could analyze each of the functions, but to observe the differences between the real-time node's callbacks and the non-real-time node, it will be sufficient to examine the time between executions and the execution time of the timer callbacks for both nodes. To view the histograms for the intervals between executions, use the following commands:

```
$ ros2 run yaets elapsed_time_histogram.py strategy_1.log --function
ConsumerNode::timer_cb --bins 40
$ ros2 run yaets elapsed_time_histogram.py strategy_1.log --function
LoggerNode::timer_cb --bins 40
```

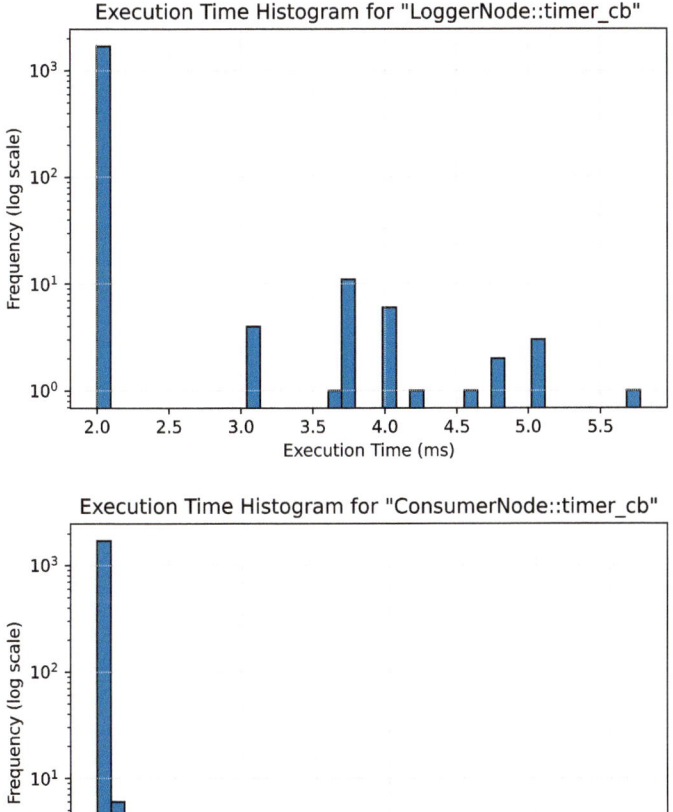

Figure 7.23: Execution time of the timer callbacks for the LoggerNode and the ConsumerNode. Again, it is important to observe the X-axis to see the differences.

Figure 7.22 shows the results of these commands. In the case of the LoggerNode, which is the timer callback of the node we did not consider critical, we can observe that, while ideally, there should be 10 milliseconds between executions to maintain a constant rate of 100 Hz, the variability is quite large, ranging from 4 milliseconds to

16 milliseconds. I want to emphasize that it may indeed reach 100 Hz, but the lack of consistency is what degrades execution from a real-time perspective. Note that some executions are delayed, while others occur earlier, because the events are triggered at the correct times but take longer to execute. This results in delayed executions overlapping with the following correct execution. This is why the histogram is nearly symmetrical. In contrast, if we observe the results of the ConsumerNode, the high-priority node, the variability is less than a few tenths of a millisecond, much smaller than in the other case, allowing for a practically constant rate, even with a heavily loaded system.

The execution time of the callbacks is also affected, degrading the response time. This is because, during the execution of these callbacks, the duration is extended as other processes are scheduled, delaying the completion of their execution. This effect can be observed by displaying the histogram of execution times with the following commands.

```
$ ros2 run yaets execution_time_histogram.py strategy_1.log --function
ConsumerNode::timer_cb --bins 40
$ ros2 run yaets execution_time_histogram.py strategy_1.log --function
LoggerNode::timer_cb --bins 40
```

The results shown in Figure 7.23 indicate that an execution that should ideally take 2 milliseconds can reach nearly 6 milliseconds in the case of the LoggerNode callback, with many readings falling outside the desired timeframe. In contrast, for the ConsumerNode, nearly all executions are around 2 milliseconds, with only one instance showing a delay close to 0.8 milliseconds. In this second case, it is possible to guarantee bounded response times very close to the ideal.

7.2.5.2 Strategy 2: Callback Groups in the Same Node with Different Priority

Putting an entire node in an executor that runs in real-time may be too coarse-grained, and we may want more selective control over the execution of nodes. A node may have events that require strict time constraints, but it may also have other events that are not as important to prioritize over those of other nodes. In this case, we can gain more control over a node's callbacks by assigning the callback groups for priority tasks directly to an executor that runs in real-time, while the remaining callback groups are placed in executors that run with normal priority, as shown in Figure 7.24.

The key changes from Strategy 1 to this approach can be summarized in the following steps:

- In the ConsumerNode, a member attribute is added to the class to store the callback group whose execution will be prioritized. Since this callback group must be accessed from main(), as we will see shortly, a public method is needed to retrieve it.

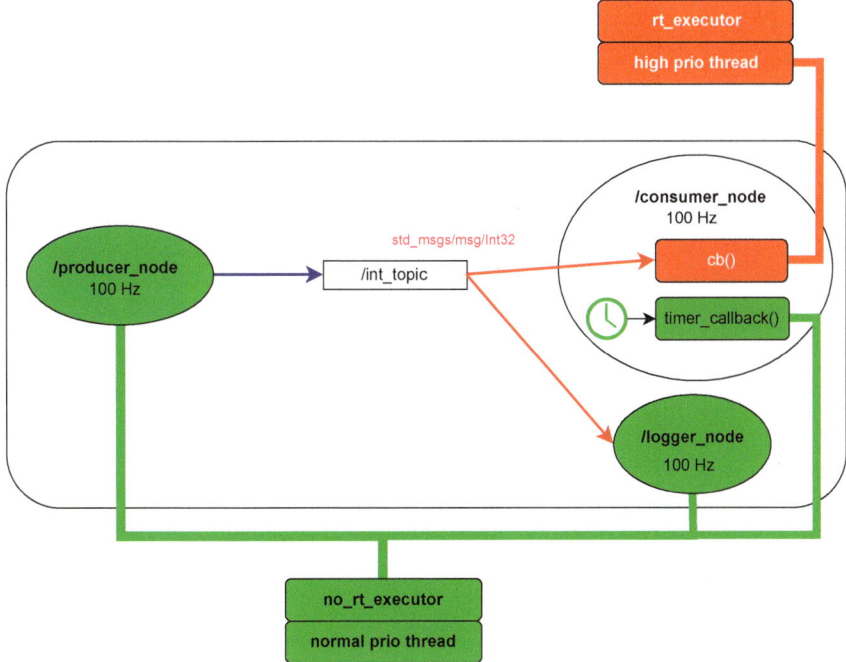

Figure 7.24: Evolution of the problem, in which different callbacks in the same node are executed in different executors.

```cpp
br2_deep_ros/src/strategy_2.cpp

class ConsumerNode : public rclcpp::Node
{
  ...
public:
  ...
  rclcpp::CallbackGroup::SharedPtr get_rt_callback_group()
  {
    return rt_callback_group_;
  }

private:
  ...
  rclcpp::CallbackGroup::SharedPtr rt_callback_group_;
```

- In the constructor, the prioritized callback group is created and used to create the subscriber that we want to execute in real-time. The other callbacks are added to the node's default callback group, as no other option is specified.

```
br2_deep_ros/src/strategy_2.cpp

ConsumerNode() : Node("consumer_node")
{
  rt_callback_group_ = this->create_callback_group(
    rclcpp::CallbackGroupType::MutuallyExclusive, false);

  rclcpp::SubscriptionOptions sub_options;
  sub_options.callback_group = rt_callback_group_;
  sub_ = create_subscription<std_msgs::msg::Int32>(
    "int_topic", 100, std::bind(&ConsumerNode::cb, this, _1), sub_options);

  timer_ = create_wall_timer(10ms, std::bind(&ConsumerNode::timer_cb, this));
}
```

The second parameter of `create_callback_group()`, set to `false`, indicates whether we want this callback group to be automatically added to an executor along with the rest of the node when the node is added to the executor.

- Finally, in main, we add all the nodes to the non-real-time executor. In the case of the `ConsumerNode`, by adding the entire node, we are actually placing the default callback group into the executor. To add the prioritized callback group to the real-time executor, we use the `add_callback_group` method, which requires the callback group and a pointer to the node that contains it.

```
br2_deep_ros/src/strategy_2.cpp

no_rt_executor.add_node(node_producer);
no_rt_executor.add_node(node_logger);
no_rt_executor.add_node(node_consumer);

rt_executor.add_callback_group(
  node_consumer->get_rt_callback_group(),
  node_consumer->get_node_base_interface());
```

Note that now the timer callback in `ConsumerNode` no longer runs in real-time, so it experiences delays in both completion and execution periodicity. In contrast, the subscription callback executes within a very tight time frame. Figure 7.25 illustrates the difference between having this callback execute in real-time versus executing like the other callbacks. In the real-time case, its execution time closely aligns with the design, with minimal variation.

As in Strategy 1, the aim of this approach is to improve the bounded update rate and reduce response time (Figure 7.26).

7.2.5.3 Strategy 3: High-Priority Callback Groups in Different Nodes

In many applications that control a robot, there are data and processing flows across different nodes that can be identified as critical. For example, if we need to react to an obstacle detected in an image, from the moment the image is captured to the generation of control commands, all callbacks in the different nodes that process the image, detection, and control—which may be part of this critical path—should execute in real-time. Placing all these nodes in a real-time executor is not suitable, as critical events would have the same priority as non-critical events. It is necessary for these callbacks to be in a prioritized callback group that runs in a real-time executor.

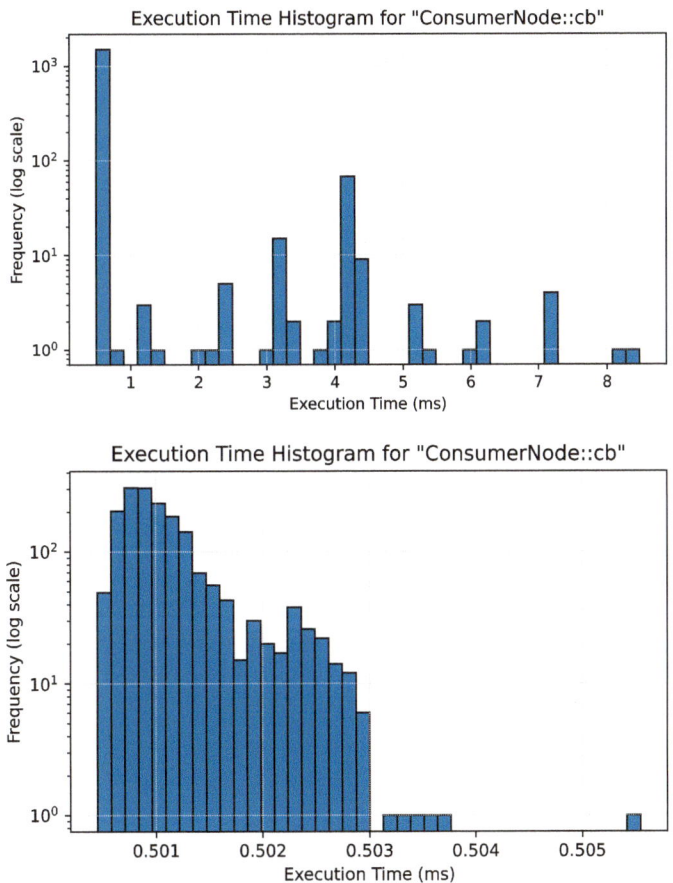

Figure 7.25: Execution time of the subscription callback for the ConsumerNode. Upper graph corresponds to non-real-time execution, and the bottom graph corresponds to real-time execution.

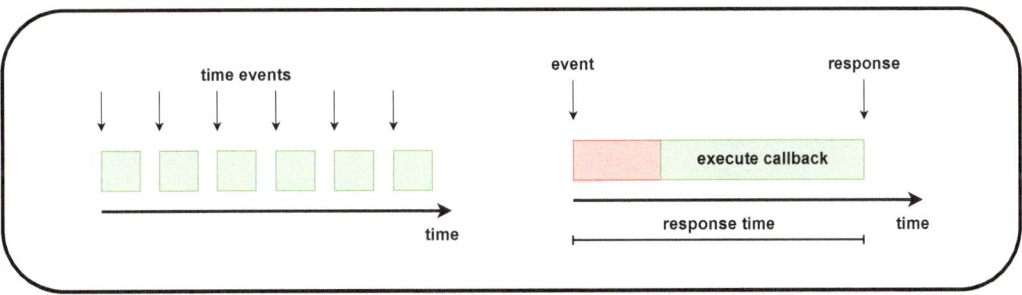

Figure 7.26: Benefits of strategy 2: Improves bounded update rate and reduces response time.

Figure 7.27 illustrates this example of a car braking system. Each node has a print_state() function that is called periodically at 10 Hz to monitor its status. Additionally, there is a logger_node that records information about the execution.

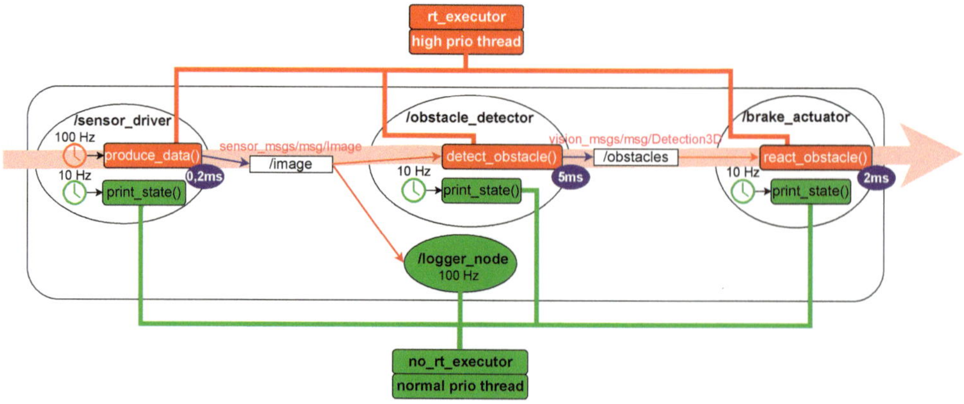

Figure 7.27: Car brake system example. The critical computation flow is marked with a big red arrow. Times of computation of critical functions are shown.

All of this is not critical and should not run in real-time, so it is placed in an executor that runs on a thread with normal priority. In the same figure, a wide, faded red arrow highlights what it is considered the application's critical processing flow, which involves image generation (**produce_data**), obstacle detection in the image (**detect_object**), and the reaction to an obstacle by sending commands to the brakes (**react_obstacle**). The goal is to ensure that the time from when the image is captured to when the brakes are activated in response to an obstacle is executed in real-time, without unbounded delays. Therefore, these three functions are placed in an executor that runs in real-time. The figure shows the time taken by each critical function, indicating that, in total, the system should apply the brakes in a little over 7 milliseconds.

- What is important to measure in this example is the reaction time from when the image is captured to when the response occurs.

```
br2_deep_ros/src/strategy_3.cpp

class SensorDriverNode : public rclcpp::Node
{
  ...
  void produce_data()
  {
    SHARED_TRACE_START("brake_process");
    waste_time(shared_from_this(), 200us);

    sensor_msgs::msg::Image image_msg;
    pub_->publish(image_msg);
  }
...
class BrakeActuatorNode : public rclcpp::Node
{
  ...
  void react_obstacle(vision_msgs::msg::Detection3D::SharedPtr msg)
  {
    waste_time(shared_from_this(), 2ms);
    SHARED_TRACE_END("brake_process");
  }
...
int main(int argc, char * argv[])
{
  ...
  SHARED_TRACE_INIT(session, "brake_process");
```

- Each of the nodes with functions in the critical path has a callback group called `rt_callback_group_` where the critical callback is added. For example, in `ObstacleDetectorNode`, there is:

```
br2_deep_ros/src/strategy_3.cpp

ObstacleDetectorNode() : Node("obstacle_detector")
{
  rt_callback_group_ = create_callback_group(
    rclcpp::CallbackGroupType::MutuallyExclusive, false);

  rclcpp::SubscriptionOptions sub_options;
  sub_options.callback_group = rt_callback_group_;

  sub_ = create_subscription<sensor_msgs::msg::Image>(
    "image", 100,
    std::bind(&ObstacleDetectorNode::detect_obstacle, this, _1),
    sub_options);

  pub_ = create_publisher<vision_msgs::msg::Detection3D>("obstacles", 100);
  timer_state_ = create_wall_timer(100ms,
    std::bind(&ObstacleDetectorNode::print_state, this));
}

rclcpp::CallbackGroup::SharedPtr get_rt_callback_group()
{
  return rt_callback_group_;
}
private:
  ...
  rclcpp::CallbackGroup::SharedPtr rt_callback_group_;
```

- Finally, in main(), we assign each node or callback group to the appropriate executor based on whether it should run on a real-time thread or not:

```
br2_deep_ros/src/strategy_3.cpp

    rclcpp::executors::SingleThreadedExecutor no_rt_executor;
    rclcpp::executors::MultiThreadedExecutor rt_executor(
      rclcpp::ExecutorOptions(), 3);

    no_rt_executor.add_node(node_sensor_driver);
    no_rt_executor.add_node(node_obstacle_detector);
    no_rt_executor.add_node(node_logger);
    no_rt_executor.add_node(node_brake_actuator);

    rt_executor.add_callback_group(
      node_sensor_driver->get_rt_callback_group(),
      node_sensor_driver->get_node_base_interface());
    rt_executor.add_callback_group(
      node_obstacle_detector->get_rt_callback_group(),
      node_obstacle_detector->get_node_base_interface());
    rt_executor.add_callback_group(
      node_brake_actuator->get_rt_callback_group(),
      node_brake_actuator->get_node_base_interface());

    auto rt_thread = std::thread(
      [&]() {
        sched_param sch;
        sch.sched_priority = 90;

        if (sched_setscheduler(0, SCHED_FIFO, &sch) == -1) {
          throw std::runtime_error{
            std::string("failed to set scheduler: ") + std::strerror(errno)};
        }

        rt_executor.spin();
    });

    no_rt_executor.spin();

    rt_thread.join();
```

It is important to understand the effects of applying these techniques in a critical system like the automatic brake control software in this example, even if it is synthetic and the execution times are artificial. This is something that can be generalized to any control program for a robot. Imagine that we do not use these real-time techniques—in other words, we do not set the scheduling algorithm to SCHED_FIFO or increase the priority to 90 in the code just above. It is easy to test this simply by commenting out the lines that set these parameters within the lambda function of rt_thread. The upper graph in Figure 7.28 shows the end-to-end times in a heavily loaded system, from obstacle perception to reaction. We observe that we often react within 20 milliseconds, but it is also common to take up to 60 milliseconds, which is unacceptable for a vehicle at certain speeds. By applying these techniques, as shown in the lower graph of the same figure, we see that the reaction time is consistently below 7.5 milliseconds, which is quite optimal, considering that the minimum reaction time, due to the duration of the callbacks in the critical path, is 7.2 milliseconds.

In this case, the beniefit is reduce End-to-end latency (Figure 7.29).

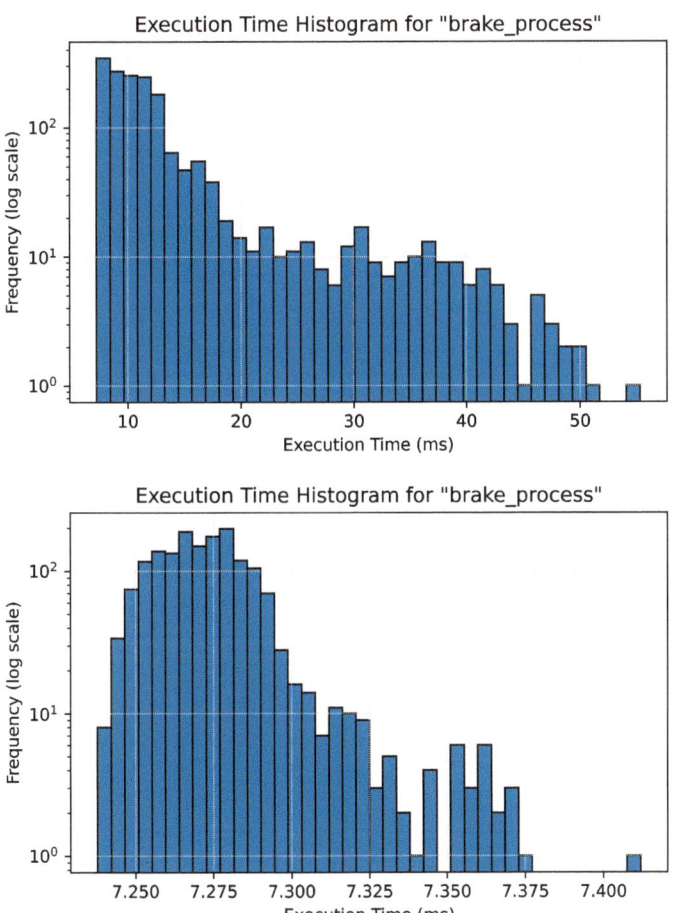

Figure 7.28: Execution time of the end-to-end processing of the critical path. Upper graph corresponds to non-real-time execution, and the bottom graph corresponds to real-time execution.

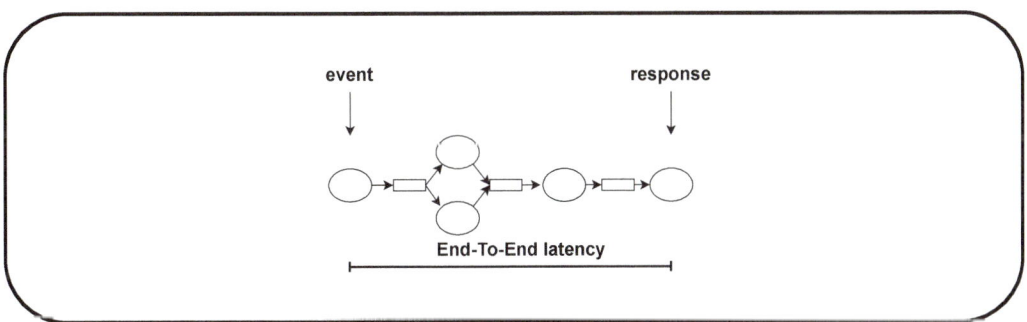

Figure 7.29: Benefits of strategy 3: End-to-end latency.

To summarize this part on Real-Time in ROS 2, the key takeaway for the reader is the importance of considering these aspects. If you have a control system that is not performing as expected, it is time to start designing your application with these real-time techniques in mind.

PROPOSED EXERCISES:

1. Design an API that allows you to explicitly set the priority for each element managed by the executor.

2. Improve the projects from previous chapters by identifying which parts should operate in real-time and applying the techniques discussed in this chapter.

## CHAPTER 8

Contributing to ROS 2 Software

I want to dedicate this final chapter to one of the aspects of ROS that has been barely covered in the previous chapters, but is what makes ROS truly great: the ROS community of users and developers. There may have been better middleware for robot programming in the past; there may be better options now, or there may be in the near future. What makes the difference is the amount of software—tools, applications, and device drivers—that you have at your disposal, and that depends solely on the size of the community. It is this that determines whether it is considered a standard, taught in universities, used in companies, and adopted by robot and device manufacturers.

On the official ROS documentation page, there is a large section dedicated to the ROS 2 Project, and within it, a part called "Contributing". The purpose of this book is not to reproduce its content, but if someone wants to actively participate in the ROS 2 Community, it is worth taking some time to read it carefully. It outlines certain procedures and conventions, as well as numerous resources, to help you get started on this exciting journey, which can lead to becoming a relevant person in the community, contributing to or even leading important projects, and having your opinion valued in technical discussions. However, I would like to reproduce here two of these principles listed in this section:

1. *Respect what came before.* ROS has been used for nearly two decades and has become a global standard in robot programming. Thousands of developers have used and contributed to ROS. Many decisions have been made through discussions involving people with substantial experience in robotics and strong technical backgrounds. If you are new to ROS, you should have the humility to consider this before jumping in to say that everything is wrong and that, luckily, you have arrived to enlighten everyone with your knowledge. It is good practice don't demand a public apology when someone points out that perhaps your comments were not accurate.

   For example, there are many people who criticize the structure of ROS, especially its build system. It is common for beginners in ROS to approach ROS

Discourse from time to time and ask, 'Why is this so complicated? Can't we just throw away colcon and ament and use basic CMake?' This is a mix of endearing ignorance and naivety. The truth is that this infrastructure is necessary to manage community contributions in the form of packages in different languages, which need to be compiled in a specific order within a workspace and be able to link at both build and runtime with all the underlays and overlays on a computer or robot. Believe me, without these tools, any attempt to create packages for others to use would require much more effort.

2. *Think about the community as a whole.* The ROS Community is very diverse, and when making decisions or implementing improvements, it is essential to consider that these need to be adaptable for people in various roles, working on very different projects, and with all kinds of robots. *ROS needs to accommodate the requirements of the entire community.*

This chapter covers topics that a ROS developer needs to know, or may not fully understand, when they consider contributing to existing ROS software or perhaps sharing their own software. It is a great idea and may be the only way to step out of your bubble and start actively participating in the ROS community. The journey can be challenging at first, but you will learn a lot along the way, and it will be very rewarding. I have colleagues at the university who are excellent professors but lack the real-world experience of testing their programs, maintaining good programming style, and empathizing with future users of their software by making it easy to use and able to justify any design decisions if questioned. Perhaps it is the fear of being questioned that keeps them in their bubble. Open source is a good cure for this.

This chapter is divided into two parts. In the first part, we will describe how to contribute to existing projects and cover some basic concepts of contributing to open-source projects. This section will cover aspects such as working with git, and the steps involved in submitting a contribution. In the second part, we will look at how to start your own project, including some decisions you need to make, such as organizing your repository into branches, choosing a license that suits your needs, setting up a continuous integration system, managing issues and pull requests, and creating binary packages for distribution from your project. I hope you enjoy it.

8.1 CONTRIBUTING TO A ROS 2 PROJECT

It is great to find your place and contribute to projects where you feel comfortable based on your experience, training, and interests. It is natural to want to contribute code, and you can find many issues labeled "help wanted" or "good first issue" in project repositories[1]. Another way to contribute is through documentation[2], which is crucial for the community. You can also assist others on Robotics Stack Exchange[3], as the legendary Gijs van der Hoorn did for years, answering hundreds, if not thousands,

---

[1]https://docs.ros.org/en/rolling/The-ROS2-Project/Contributing.html#what-to-work-on

[2]https://docs.ros.org/en/rolling/The-ROS2-Project/Contributing/Contributing-To-ROS-2-Documentation.html#

[3]https://robotics.stackexchange.com/

of questions on the old ROS Answers, helping many ROS developers in need, earning my respect forever.

However, we are going to focus on the first form of contribution, which is the most relevant for this book's approach: contributing code. It is probably most natural to start by contributing to the projects you actually use in your daily work, perhaps because you have found a bug and submitted an issue (this is already a contribution!). Personally, whenever I can, I like to include a potential solution for each issue, which I then materialize in a contribution to the repo. Or maybe you think of an improvement, possibly one that benefits you directly because you need it for the tasks you are working on. If this is not the case, I recommend some core projects that, due to their respected members, can teach you good practices and strategies for how to do things within ROS:

- **Nav2** (`https://github.com/ros2/rviz`). This is the project from which I have learned the most by contributing to it. Even when I have a question about how to do something, I go to this repo to see how Steve Macenski, one of the most respected ROS programmers, has done it. There is a Navigation Working Group that meets periodically. If you are interested in robot navigation, drop by someday.

- **MoveIt2** (`https://github.com/moveit/moveit2`). If, on the other hand, you are interested in or work with robotic arm manipulation, this is your project. It is also related to the Manipulation Working Group.

- **ros2_control** (`https://github.com/ros-controls/ros2_control`). This is a framework for real-time control of robots. It aims to standardize the way commands are sent to the robot's actuators and allows you to use a wide range of algorithms or test your own. Bence Magyar is the main maintainer of this repository and also coordinates the ros_control Working Group.

- **rclcpp** (`https://github.com/ros2/rclcpp`). If you enjoy programming in C++ and found the content in chapter 7 interesting, this is a great place to contribute. Few things are as core as the C++ client libraries.

- **rcl_rs** (`https://github.com/ros2-rust/ros2_rust`). If you enjoy programming in Rust and therefore have plans for everyone to transition their projects to this language, this is the place for you. There is a growing interest in Rust playing a significant role in ROS 2 in the near future. It will not be easy, but they are looking for brave contributors here.

- **Gazebo** (`https://github.com/gazebosim`). Gazebo is another major open-source project within ROS. There are other highly respected simulators like Unity, Open 3D Engine (O3DE), or Webots—whose journey from proprietary to open-source software I want to highlight—but Gazebo has always been a part of ROS, and it has something special.

- **RViz** (`https://github.com/ros2/rviz`). This tool, which we have been using since the beginning of the book, is one of the treasures of ROS. Visualization tools are extremely important in robotics, and this one has fulfilled its mission. Recently, alternatives like Foxglove have appeared, which are very attractive and which I also recommend using, but I believe a free alternative should always exist for this type of tool.

- **PlanSys2** (`https://github.com/PlanSys2/ros2\_planning\_system`). Okay, I know this project is mine, and that is why I am including it, and for that, I apologize. But it is a great framework for symbolic AI planning, and I also believe it serves as a reference for good programming practices.

As you can see, with a few exceptions—which, by the way, I find very healthy in terms of having more alternatives—GitHub is the main platform where ROS developers host their projects. For these reasons, this book focuses on using git and on GitHub's interface and features. In the first chapter, I mentioned that I assume the reader is an experienced programmer, but I was too when I started with ROS, and I had to learn that using git involved more than the sequence of `git add`, `git commit`, and `git push` to my repository. So let us start there, by discussing how to use git with repositories that are not your own. I have encountered this with users wanting to contribute to my projects who, when they had something ready, would ask me for permissions on my repository. This is not typical in an open-source project, and ROS is no exception. You need to learn how to make *Pull Requests*. I have created a git cheatsheet (Figure 8.21) at the end of this chapter with the commands I commonly use when working with my repositories, and a diagram that is often missing in those I have found online, which usually do not include how to work with different repositories: the original one you want to contribute to and the one you created by forking the original. You can find this cheatsheet at the end of this chapter.

A contribution can primarily be an improvement to the documentation, fixing a bug, or adding a new feature. You can also bitterly complain about something, but this rarely leads to anything productive. To illustrate the concepts and steps for contributing, I will use some of my repositories, included the one of this book, which contains several packages with GitHub Actions for continuous integration that run unit tests and style checks for each package.

I recommend discussing your plans in an issue before starting to work on something, just in case you end up working in vain. Through an issue, you can get feedback directly from the maintainers about whether something is truly a bug, or if the new feature you are proposing is not of interest, or if someone else is already working on it.

Typically, many maintainers have created templates for reporting issues. If one exists, use it. Otherwise, your issue may never be addressed. At the time of writing this book (Autum, 2024), there is a very interesting and illustrative issue at `https://github.com/fmrico/book_ros2/issues/30` open by one of my students in the Bachelor's Degree in Software Robotics Engineering at Universidad Rey Juan Carlos in Madrid. It seems he has found this the perfect moment to exact his revenge after years of classes, exercises, and exams on my part. He has correctly used the template for requesting a change on this book.

A slightly more technical issue, reporting a bug, can be found in the PlanSys2 repository on `https://github.com/PlanSys2/ros2_planning_system/issues/329`. It is a minor bug, but it is interesting because I, as the principal author of PlanSys2, raised a question about a bug related to an error message that does not seem to be an actual error and appears persistently on the screen when this software is executed. After a couple of messages, including input from the author of this part of the code (identified using `git blame`), we concluded that it is indeed a bug and that the most appropriate solution is to remove this error message.

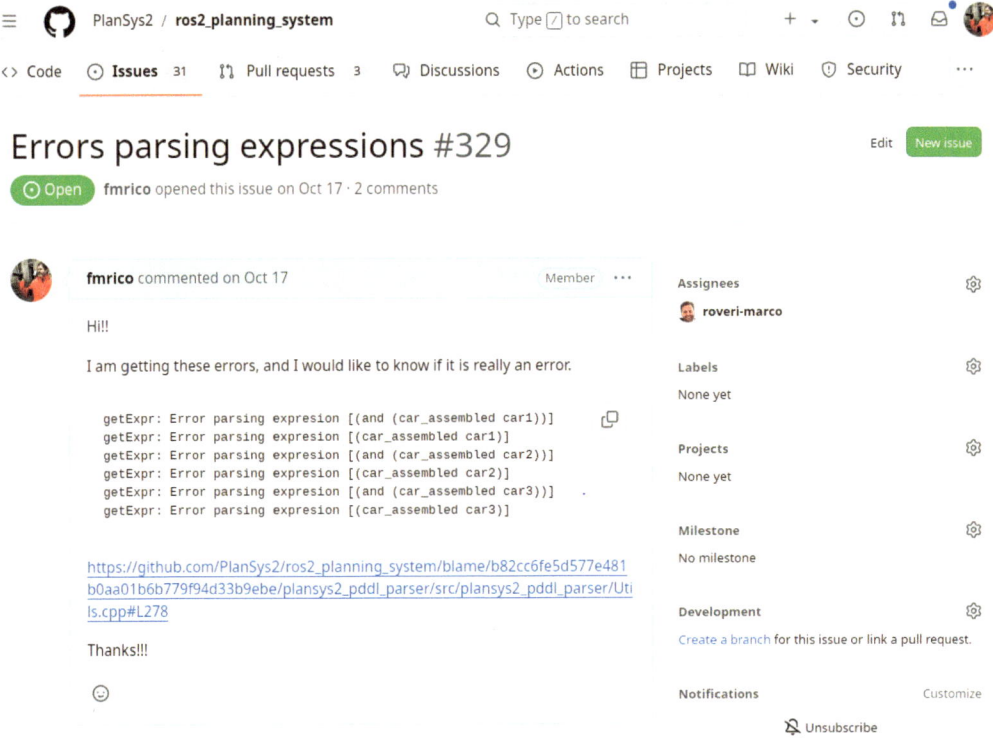

Figure 8.1: Technical issue to fix a bug.

Even though I am the principal maintainer of this software, I will proceed as if I were a regular user wanting to contribute a solution to this bug without permissions in this repo. It is likely that before creating this issue, I would have cloned this repository into my workspace, installed the dependencies, and compiled it from these sources:

```
$ mkdir -p planning_ws/src
$ cd planning_ws/src
$ git clone https://github.com/PlanSys2/ros2_planning_system.git
$ vcs-import . < ros2_planning_system/dependency_repos.repos
$ cd ..
$ rosdep install --from-paths src --ignore-src -r
$ colcon build --symlink-install
```

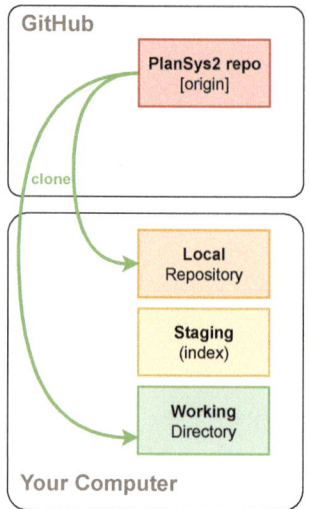

Figure 8.2: Diagrams after clonning the reposotory into the local computer.

The third command has created the situation shown in Figure 8.2. As you should know if you have worked with Git before, a local repository is the complete copy of a project stored on your machine, including both the project files and the repository's change history and metadata. The staging area, on the other hand, is an intermediate zone where changes are prepared before committing them, allowing you to select which modifications will be included in the repository's history. Finally, the working directory is the space where you edit the project's files and reflects the most recent version of the repository, unless there are uncommitted local changes.

You can verify that your local repository points to the remote repository, which has been identified as `origin`:

```
$ cd planning_ws/src/ros2_planning_system
(rolling)ᵃ$ git remote show
origin
(rolling)$ git remote show origin
* remote origin
 Fetch URL: https://github.com/PlanSys2/ros2_planning_system.git
 Push URL: https://github.com/PlanSys2/ros2_planning_system.git
 HEAD branch: rolling
 Remote branches:
 [...]
```

  ᵃFrom this point forward, whenever we are in a Git repository's working directory, the current branch will be displayed in parentheses before the $ character.

The first step is to fork the PlanSys2 repository into the user's account, in my case at `https://github.com/fmrico/ros2_planning_system`, pushing the button

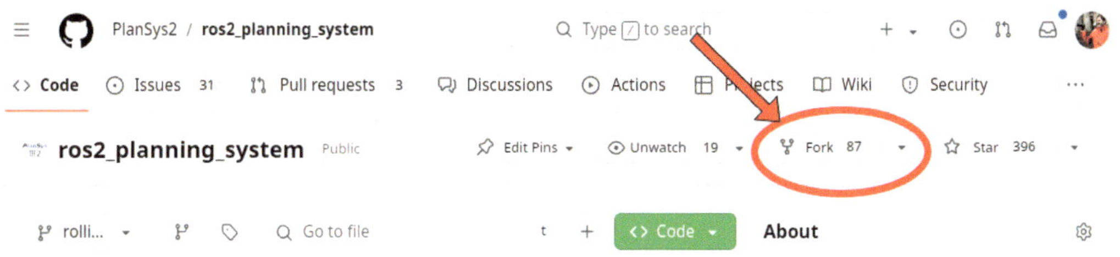

Figure 8.3: Press this button to fork this repository in your user account.

shown in Figure 8.3. Next, I will add a new remote on my local machine to work with this repository. By default, the identifier for the remote of a normal clone is `origin`. When adding a new remote, we can assign it an identifier to refer to it. Many people call this second remote `upstream`, but I prefer to name it in a way that helps me remember where it originates from, as I might be working with multiple remotes. I will name this new remote `fmrico`. With the following command, I will have the situation shown in Figure 8.4:

```
(rolling)$ git remote add fmrico https://github.com/fmrico/ros2_planning_system
```

We can use the commands `git remote show` and `git remote show fmrico` to verify that a new remote exists and that it points to the correct location.

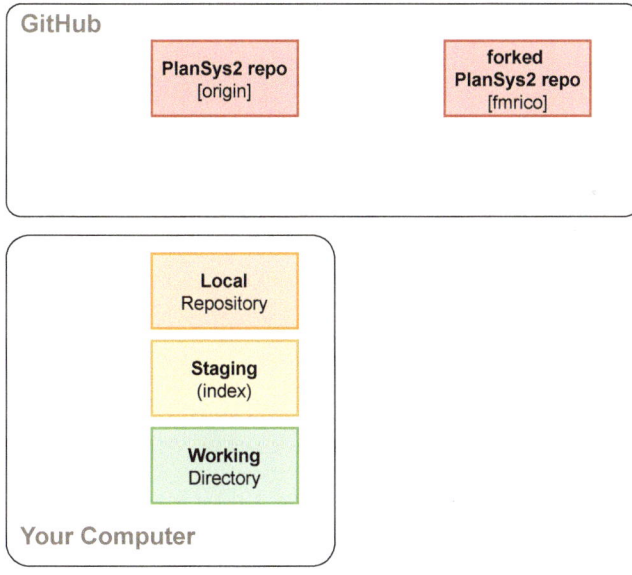

Figure 8.4: Diagrams after forking the original repository.

Every contribution should be made in an independent branch that starts from the branch to which you want to contribute. We were already in the 'rolling'' branch, and now we will create a branch that we will call, for example, `fix_issue_329`:

```
(rolling)$ git checkout -b fix_issue_329
(fix_issue_329)$
```

With this last command, we create the branch and switch to it. Now we can start working on fixing this bug. Once we have finished, which includes compiling, testing, and passing all tests, we stage the files we have changed and commit them:

```
(fix_issue_329)$ git add plansys2_pddl_parser/src/plansys2_pddl_parser/Utils.cpp
(fix_issue_329)$ git commit -s -m "Fixing bug #329"
```

The `-s` option in the commit is because all commits must be signed.

Finally, we push the branch to the `fmrico` remote, which is assumed to be the only remote where we have permission to upload changes, as it is our own repository:

```
(fix_issue_329)$ git push --set-upstream fmrico fix_issue_329
```

The last command provides you with a URL to directly create the pull request, or you can go to the repository `https://github.com/fmrico/ros2_planning_system` and do it from there. In any case, it is important to verify that the target branch is correct and to provide a good description. At the very least, reference the issue being addressed, as shown in Figure 8.5.

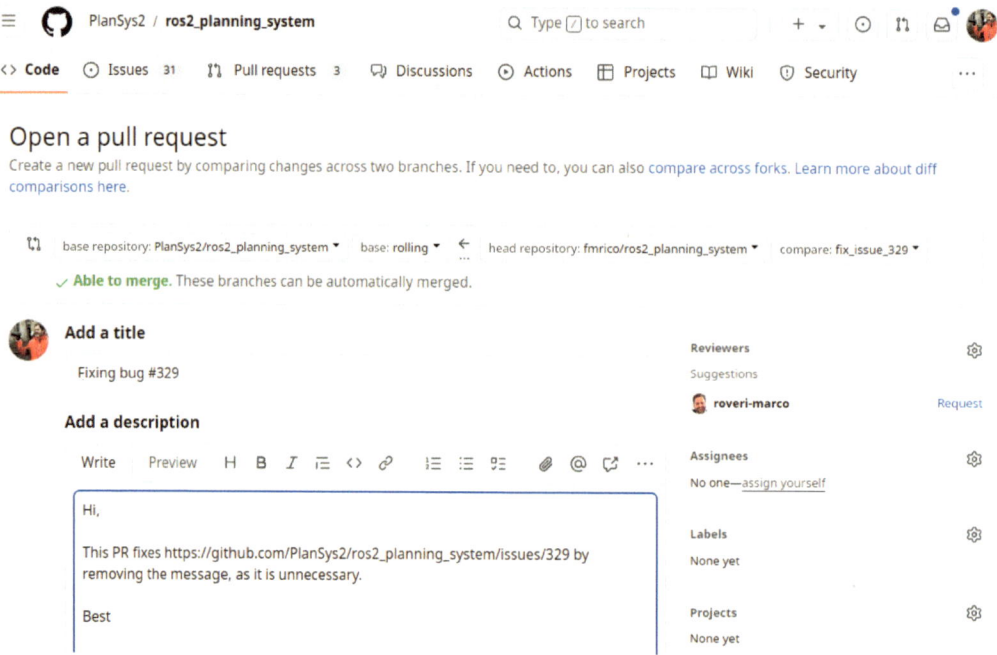

Figure 8.5: Writing the pull request to fix the issue.

Recapping, with the last few commands, once we have fixed the bug on our local machine, verified that everything works, and passed the tests, we have followed the workflow shown in Figure 8.6. Now we just need to wait for the repository maintainer to provide feedback. During this process, any commits we make to address this feedback will automatically be added to the pull request when we push.

During this process, new changes might be incorporated into the original repository, which could include resolving conflicts with new commits or performing a merge/rebase to account for them in your pull request.

Once everything is ready and the maintainer deems it appropriate, they will perform a merge, closing the pull request and incorporating our contribution into the repository. If we were not already contributors to PlanSys2, we now are. At this point, we can delete the branch `fix_issue_329`, switch back to the `rolling` branch, and perform a pull to receive the newly incorporated changes.

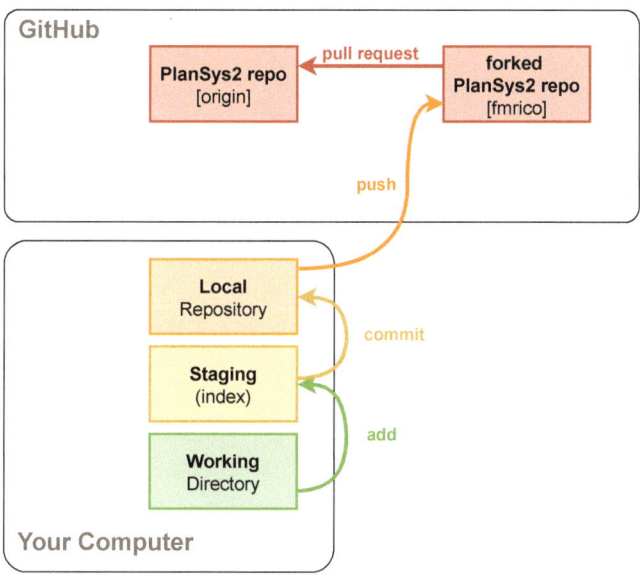

Figure 8.6: Itinerary of the contribution from our local machine to the `origin` repository.

8.2 CREATING AND MANAGING YOUR OWN ROS 2 PROJECT

ROS 2 has been an opportunity for those who did not join the ROS 1 ecosystem from the beginning and arrived to find that there were already packages, libraries, and frameworks for almost everything. Of course, this situation is also desirable, and contributing to existing packages is of great value and can lead to having significant influence over the repository you contribute to. For instance, I am considering passing the maintenance of PlanSys2 to a reliable contributor eager to revitalize its development. However, it is natural to want to lead projects—though one should never fall into the trap of reinventing the wheel.

When ROS 2 started, for example, there was a need for a navigation framework to inherit the success of the ROS 1 navigation stack. Steve Macenski seized the opportunity to create an improved reimplementation of this navigation framework, Nav2, becoming one of the most prominent programmers in the ROS Community.

He later founded Open Navigation, a company that will surely help him buy his next Ferrari to drive to his luxurious yacht. It was then that I realized that ROSPlan, the leading PDDL-based symbolic planning framework in ROS 1, had no plans (nor does it yet) to migrate to ROS 2—a task I would have gladly contributed to. For this reason, I started developing PlanSys2 for ROS 2, significantly improving on its predecessor. This has not brought me infinite wealth, but it has bring me many benefits. It is also deeply satisfying to see it being used in companies, in universities as the foundation for doctoral theses, or that PlanSys2 is now on the Astrobee robots that NASA has on the International Space Station.

This pattern can be seen across many reference packages, libraries, and frameworks in ROS 1. ROS 2 reshuffled the deck, giving newcomers to the ROS Community the chance to take the lead on reference packages. This window of opportunity still exists. While some gaps have been filled, advancements in robotics—such as the rise of LLMs (Large Language Models) and their applications in robots—continue to open new opportunities to be seized, furthering the glory of ROS.

Creating and maintaining a software package, especially if it achieves some success and is used by others, with users and contributors depending on it, is not easy. It requires a level of responsibility to keep maintaining it, even years after you created it, and perhaps when it is no longer your active area of focus. People need your attention to review issues and pull requests, and you feel a certain commitment to keep generating binary packages for each ROS 2 distribution. In this section, we will address some of the essential aspects of this process, keeping in mind that every project has its own idiosyncrasies, requirements, and objectives. What I will share here is what I consider relevant after several years of maintaining close to a dozen repositories that have achieved a certain level of adoption within the ROS Community. Specifically, I want to address the following topics:

1. Licenses and open-source business models.

2. The ROS development model based on distributions.

3. Coontinuous Integration (CI).

4. Repository Structure and Documentation.

5. Binary packages creation.

6. Tips and best practices for managing contributions.

Let's get started!

8.2.1 Licenses and Open-Source Business Models

A software license is a legal document that defines how users can use, modify, and redistribute a computer program. The license defines which rights copyright holders reserve for themselves, and which ones are granted to those receiving the software. Licenses can be open source, promoting collaboration and access to the source code, or proprietary, restricting usage and modification. Licenses are important because they provide legal clarity, prevent misuse of the software, and allow developers to control how their work is used.

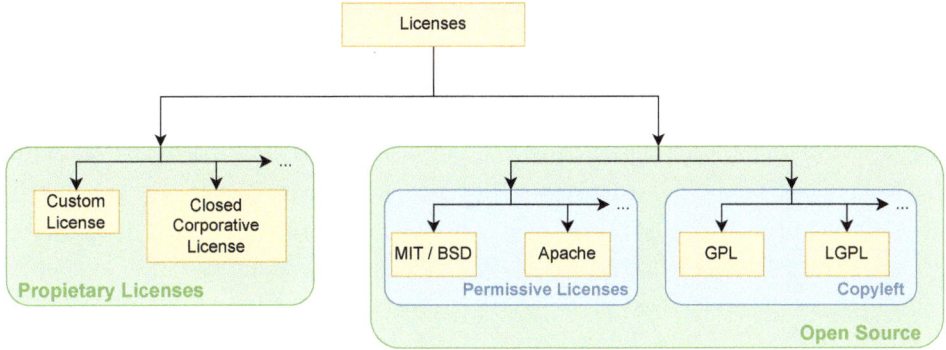

Figure 8.7: Software licenses classification.

The term open source does not merely refer to making source code available. The open-source Initiative (OSI) Definition[4] encompasses aspects such as free redistribution, allowing derivative works, ensuring the integrity of the author's source code, and prohibiting discrimination against certain uses or users, among other principles. This definition is grounded in early definitions of *free software,* notably the Debian Free Software Guidelines[5], which themselves draw inspiration from foundational documents by the Free Software Foundation, such as "What is free software"[6].

Let me describe the main differences between software licenses using the classification shown in the diagram in Figure 8.7. The first major division is between proprietary and open-source licenses. The fundamental difference is that open-source licenses guarantee the four freedoms of open-source software: the autonomy of users to use, study, share, and modify the software[7].

These freedoms include the ability to run the software for any purpose without restrictions on its use, whether personal, educational, commercial, or social. They also grant the right to study how the software works by analyzing and modifying its source code to adapt it to specific needs. Additionally, users have the freedom to redistribute copies of the software, whether in its original or modified form, fostering collaboration and sharing. Finally, these licenses allow users to improve the software and share those improvements with the community, promoting the evolution and collective progress of the software.

If you are thinking of including clauses in your license, such as prohibiting your arch nemesis from using your library, preventing Buddhists from using it, or banning

---

[4]https://opensource.org/osd

[5]https://wiki.debian.org/DebianFreeSoftwareGuidelines

[6]https://www.gnu.org/philosophy/free-sw.en.html

[7]Those four freedoms were operationalized by the Debian Free Software Guidelines (by setting a checklist that free software should fulfill), and that they in turn were the basis for the open-source definition.

its use in killer robots, go ahead, but it will not be open source according to the definition outlined in the paragraphs above. This would fall under the category of custom proprietary licenses[8]. If you choose open source, with certain nuances, restrictions, and obligations that we will discuss shortly, it is akin to leaving your software on the doorstep for anyone to use. You have made it free, with all the consequences that entails.

If you choose an open-source license, you can opt for either copyleft or permissive licenses. The main difference is that copyleft licenses require any derivative work to maintain the same rights if the software is distributed to other parties. For example, if you license your software with the GPL, you are requiring that, if any derivative work is distributed, it must also be distributed under the GPL. For this reason, these licenses are often referred to as "viral", as they mandate that any software incorporating them must also be distributed under the GPL.

Two examples are the Qt libraries and the MySQL database: any software that includes either of these components is required to be GPL. At that point, you can either comply and make everything GPL—like some Samsung appliances that include a sticker with a URL to request the source code for your dishwasher—or you can refuse. In the latter case, you can negotiate with the Qt developers to allow you to use their library under a different license that permits closed-source software, usually for a considerable fee. This is a business model around open-source software that many people choose. For it to happen, authors must retain the copyright.

The difference between GPL (General Public License) and LGPL (Lesser General Public License) is that while GPL requires all software it interacts with, including dynamically linked software, to comply with the conditions of the license—such as making the source code open—the LGPL does not require the dynamically linked software to adhere to the same license. For this reason, the 'L' in LGPL originally stood for 'Library'.

Merely integrating multiple components within the same system does not necessarily imply that everything constitutes a "derivative work" of any individual component. For instance, it is entirely permissible to distribute Linux, which is GPL, alongside proprietary software (e.g., within a Docker container image or on a traditional CD), and this does not obligate the proprietary software to comply with the GPL.

Permissive licenses have minimal restrictions and do not aim to ensure that modified versions of the software remain free and publicly available. Generally, they only require that the original copyright notice is retained. As a result, derivative works or future versions of software under a permissive license can be released as proprietary software. Many people choose this type of license because it removes objections that individuals or companies might have about using your software, encouraging its adoption by a larger number of users. It also happens that the original copyright holder, or in general anyone receiving the software, don't have access to the modifications

---

[8]There has been an ongoing debate since the 1990s, if not earlier, about whether some of these would fall under what are considered ethical licenses or licenses aimed at preventing large companies from benefiting from free software without contributing back. These licenses are not (clearly) free software or open-source software, but in some cases, they attempt to present themselves as such.

and improvements performed on the software, nor permission to work on them or redistribute them.

The MIT and BSD licenses, originating from the academic world, operate under the principle that the software has already been paid for by taxpayers. Therefore, their motivation is to contribute knowledge to society as freely as possible. The MIT license, for example, is very simple and can literally be included here:

> *Copyright YEAR COPYRIGHT HOLDER*
> *Permission is hereby granted, free of charge, to any person obtaining a copy of this software and associated documentation files (the "Software"), to deal in the Software without restriction, including without limitation the rights to use, copy, modify, merge, publish, distribute, sublicense, and/or sell copies of the Software, and to permit persons to whom the Software is furnished to do so, subject to the following conditions:*
> *The above copyright notice and this permission notice shall be included in all copies or substantial portions of the Software.*
> *THE SOFTWARE IS PROVIDED "AS IS", WITHOUT WARRANTY OF ANY KIND, EXPRESS OR IMPLIED, INCLUDING BUT NOT LIMITED TO THE WARRANTIES OF MERCHANTABILITY, FITNESS FOR A PARTICULAR PURPOSE AND NONINFRINGEMENT. IN NO EVENT SHALL THE AUTHORS OR COPYRIGHT HOLDERS BE LIABLE FOR ANY CLAIM, DAMAGES OR OTHER LIABILITY, WHETHER IN AN ACTION OF CONTRACT, TORT OR OTHERWISE, ARISING FROM, OUT OF OR IN CONNECTION WITH THE SOFTWARE OR THE USE OR OTHER DEALINGS IN THE SOFTWARE.*

The Apache license is somewhat more complex, detailing aspects such as software patents. It provides, for example, stronger protections for developers and users, preventing potential patent lawsuits from contributors or licensees.

If you hold the full copyright over a piece of software, you can change its license without any issues. Normally, when someone contributes to a software project, they become copyright holders, in combination with other previous copyright holders, of the software, often explicitly noted in the licenses included at the beginning of the source files they have contributed to. This means that the original author of the software cannot change the license after receiving contributions unless they follow a practice like that of the Free Software Foundation, which explicitly requires a signed copyright waiver for each contribution.

Figure 8.8 shows a graph illustrating the number of packages in the Noetic (ROS 1) and Jazzy (ROS 2) distributions that chose one license or another. It is evident that permissive licenses are the clear winners, encouraging ROS to be used with minimal restrictions by users and companies, thereby promoting its widespread adoption.

Regarding the business models offered by open source, let me introduce some of them:

1. **Consulting and Technical Support.** Developers or companies provide technical support, consulting, or specialized training on the software. For instance,

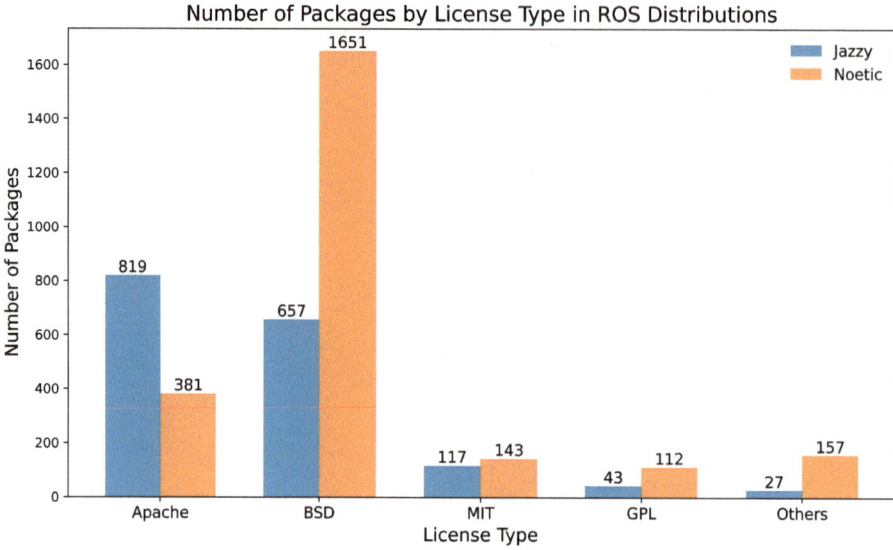

Figure 8.8: Main licenses used in packages for Noetic (ROS 1) and Jazzy (ROS 2) distributions.

companies assist with configuring and customizing ROS for industrial or research robots. This model monetizes expert knowledge without closing the code.

2. **Dual Licensing.** Software is offered under an open-source license but also with a proprietary license option for clients requiring specific features, additional support, or alternative rights (e.g., avoiding copyleft). We have already introduced the case of Qt and MySQL. In the ROS ecosystem, while most components use permissive licenses like Apache 2.0 or BSD, some libraries or extensions might be distributed under dual licensing. This provides revenue for companies developing critical or specialized tools.

3. **Commercial Distributions.** Companies package and distribute optimized, certified, or customized versions of the software, often with additional features or exclusive integrations. In ROS, companies like Canonical or Open Robotics offer certified ROS distributions, such as ROS 2 tailored for industrial applications. This adds value by providing stability and compatibility guarantees.

4. **Hybrid Products (Hardware + Software).** Open-source software is bundled with hardware products, such as robots, drones, or research platforms. For example, robots like TurtleBot rely on ROS as the foundation for their software, while the primary business involves selling hardware. ROS drives hardware adoption by being familiar and extensible.

5. **Proprietary Extensions and Plugins.** Companies develop proprietary add-ons that enhance the capabilities of open-source software. In the ROS ecosystem, this includes advanced control plugins, exclusive simulation tools for

Gazebo, or proprietary solutions for navigation or SLAM. Users pay for specific features not available in the free version.

6. **Cloud Infrastructure.** This business model is based on companies offering cloud-based services to run or integrate open-source software, often through subscriptions. For example, AWS RoboMaker provides cloud simulation and deployment services designed for ROS, combining the ROS environment with scalable computing infrastructure. This ensures a steady revenue stream through subscriptions.

7. **Certification and Training.** Some companies and organizations offer official training and certification programs for the software, ensuring a base of skilled users. For example, ROS certifications (such as those offered by The Construct) are popular within the community. This approach monetizes training without restricting the adoption of the software.

8. **Community-Based Funding.** Developers or organizations receive support through donations, sponsorships, or crowdfunding. Foundations like Open Robotics have relied for many years on support from companies and government agencies. This allows development to proceed without being tied to commercial interests.

8.2.2 The ROS Development Model Based on Distributions

In ROS 2, you develop software for a specific ROS distribution, which, let us not forget, is tied to a particular version of Ubuntu for those developing on Linux—something we agreed is the reference operating system for this book (at least until Microsoft or someone at Apple funds the third edition of this book to broaden its horizons). This has a certain impact on the versions of libraries and programming languages you can use. Just ask anyone who lived through the traumatic transition from Galactic (Ubuntu 20.04) to Humble (22.04), which forced the use of Python 3—or myself, having to migrate the simulated robot to Gazebo Harmonic because Gazebo Classic is no longer supported in Jazzy.

Here, you cannot force your users to use an older or newer version of a library: you must use the version that comes with your distribution. Period. When we participated in one RoboCup some time ago, a developer from another university who was part of our team used the latest versions of the library PCL to program their perception module. When the time came to integrate it, the fact that the rest of the team had to manually compile a specific version of PCL that was not available as a binary package led to their work being thrown in the trash. Genius.

My recommendation, which aligns with what is being done in the core ROS 2 packages, is to always develop new features using the Rolling distribution. In fact, your default branch on GitHub should be `rolling` instead of `main` or `master`. I understand that this exposes you—albeit increasingly less often—to occasional breaks due to changes in the API of a package you are using or requiring you to develop against the source code of another package that does not provide binary packages for

Rolling, as is the case with Nav2. However, the advantages outweigh the drawbacks: First, as a developer, you will always be up to date with the latest developments, at least in the core ROS 2 packages. Secondly, another advantage is the ease of creating versions for upcoming distributions. When a new distro is released—every year on World Turtle Day—that distro is generated from Rolling (upper part of Figure 8.9), which was frozen a few months prior. If you follow the same approach by creating a branch in your repository for the new distribution from the rolling branch, it is a nearly seamless process. This puts you in a position to generate binary packages for the new distribution (which we will cover in an upcoming section) immediately.

Once you have created a branch for a specific distribution, you no longer add new functionality or change any APIs in that branch; you only fix bugs. Companies and organizations require stable versions when using your software for their products, and this provides the stability they need by relying on a stable branch.

On the bottom part of Figure 8.9, you can see a repository that follows this principle. Its main branch is rolling, and it has branches like `iron-devel`, `jazzy-devel`, etc., which correspond to the branches used to produce versions for each distribution. These branches are also the source for generating binary packages.

8.2.3 Continuous Integration for Your Repository

I do not intend to oversimplify a field as broad and complex as Continuous Integration (CI) into just a few lines, risking Ruffin White—the most knowledgeable person I know in this field in the ROS Community—showing up at the next ROSCon to demand an explanation. My aim is simply to provide a straightforward solution to introduce the reader to this field and quickly equip their repository with a useful mechanism to verify that its state and the contributions it receives at least compile on a freshly installed machine with ROS 2 and the necessary dependencies, and that it passes all tests, both style checks and those programmed for each package. If the reader delves deeper into this field, they will discover that they can automatically generate Docker images with their software, create binary packages, test their software automatically in a simulator, and much more.

The simplest way to implement CI in your repository is by using GitHub Actions. All you need is to add a file for each CI action you want to configure. Let us look at how this has been done in the repository for this book, which is available at `https://github.com/fmrico/book_ros2`. The relevant structure of the repository for this explanation is shown in the following diagram:

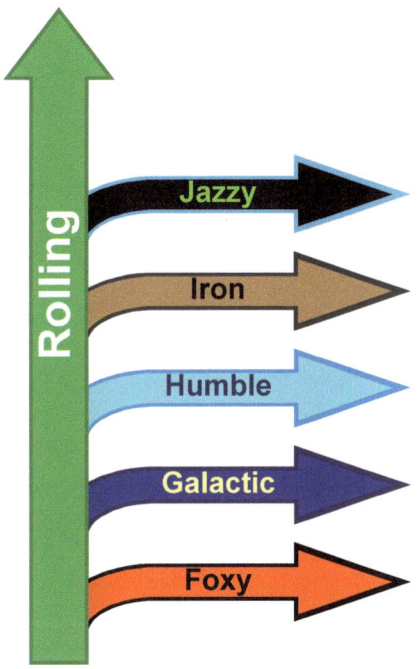

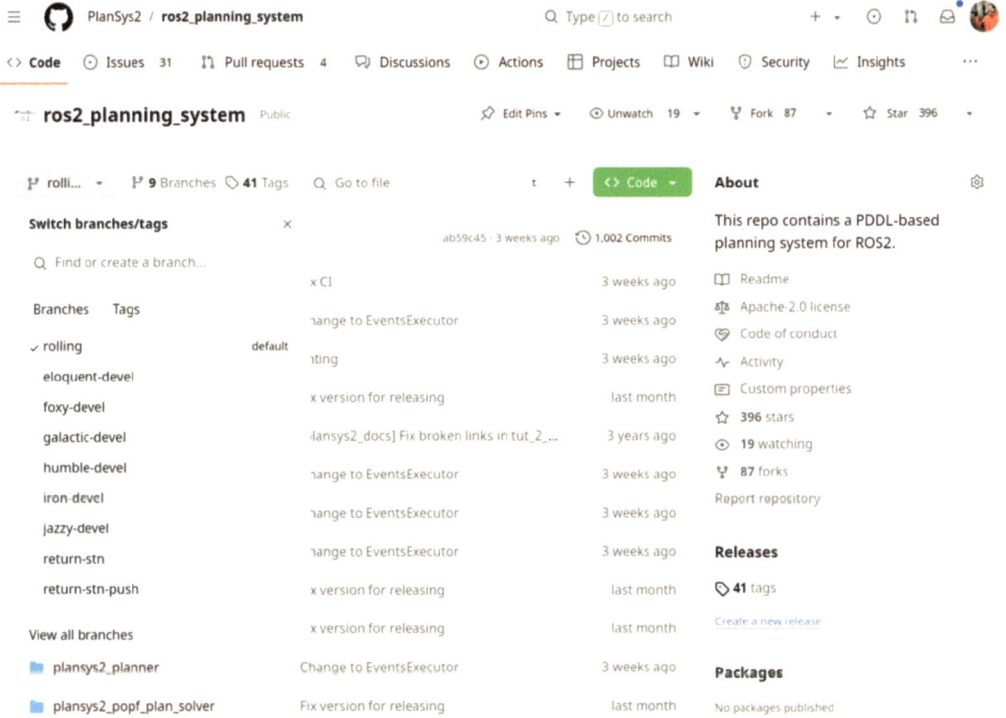

Figure 8.9: All distributions part from Rolling (upper). Repository that follows a development based on distributions (bottom).

```
Repository at https://github.com/fmrico/book_ros2

├── .github
│   ├── ISSUE_TEMPLATE
│   └── workflows
│       ├── rolling.yaml
│       ├── foxy-devel.yaml
│       ├── humble-devel.yaml
│       └── jazzy-devel.yaml
├── br_basics
├── br_bt_bump_go
└── [...]
```

Each of these files represents a new CI action that triggers workflows, which are the executions of the CI action. The file can have any name, as the content is what matters, but in a basic configuration, it is recommended to have one file per branch, as we are following the distribution-based development model outlined in the previous section. Let me analyze the contents of the rolling.yaml file, piece by piece:

- **name:** The name of the workflow.

```
.github/workflows/rolling.yaml

name: rolling
```

- **on:** Specify the cases in which this workflow will be triggered. In this case, it is set to trigger when there are pull requests or direct pushes to the rolling branch.

```
.github/workflows/rolling.yaml

on:
  pull_request:
    branches:
      - rolling
  push:
    branches:
      - rolling
```

If you are like me and enjoy waking up early on Saturdays to a fresh email in case the CI fails, you can schedule it cron-style. Ah, I love the smell of CI failures early in the morning.

```
schedule:
  - cron: '0 0 * * 6'
```

- **jobs:** Starting from here begins the specification of the jobs that will be executed, each one with a job_id.

```
.github/workflows/rolling.yaml

jobs:
  build-and-test:
    runs-on: ${{ matrix.os }}
    strategy:
      matrix:
        os: [ubuntu-24.04]
      fail-fast: false
```

For historical reasons that I will not detail, I use a `jobs.<job_id>.strategy` label, which would allow multiple jobs to run in parallel for each combination of the different values listed under this label. In reality, there is only one label (`jobs.<job_id>.strategy.matrix.os`) under `jobs.<job_id>.strategy.matrix`, and it is a single-item list: `ubuntu-24.04`. This means there will only be one execution, and the `jobs.<job_id>.strategy. fail-fast` label—indicating whether all jobs should be immediately canceled as soon as one fails—will not have any effect.

The label `jobs.<job_id>.runs-on` defines the type of machine to run the job on. `ubuntu-24.04`[9] is a valid id for a machine running an standard Ubuntu 24.04 LTS system.

- **`jobs.<job_id>.steps:`** For each job, the list of steps to be performed in this workflow is specified next. Each step can have a name. To define what is done in each step, you can directly specify a command using the run label, or you can use an action available on GitHub.

```
.github/workflows/rolling.yaml

- name: Install popf deps
  run: sudo apt-get install libfl-dev
- uses: actions/checkout@v2
- name: Setup ROS 2
  uses: ros-tooling/setup-ros@0.7.9
  with:
    required-ros-distributions: rolling
```

The first action (`Install popf deps`) installs a dependency that is needed[10] by calling `apt-get install`. Keep in mind that this is like installing a fresh, standard Ubuntu 24.04 LTS, which only includes basic packages—no ROS or anything else. This is precisely what we want.

Often, we do not specify dependencies correctly in our `package.xml` files, which would otherwise be installed with a `rosdep` call. Our project compiles because we likely installed those dependencies earlier. However, a new user downloading your project for the first time, running `rosdep`, will encounter compilation errors because you forgot some dependencies. And no, 'it works on my machine' is not a valid response to that user. This is the real utility of CI: starting with

---

[9]https://docs.github.com/en/actions/writing-workflows/workflow-syntax-for-github-actions#standard-github-hosted-runners-for-public-repositories

[10]In fact, I believe this is unnecessary and probably slipped in from another file in a different repository, but it helps me explain the different options.

clean machines every time to compile and test your project„ detecting errors that others might encounter.

The second action, with no name, directly uses the action `actions/checkout@v2`, whose URL you can easily deduce (`<repo>@<release>`) and is available at `https://github.com/actions/checkout`. This action downloads your repository into the `$GITHUB_WORKSPACE` directory for building and testing. While this action might not be strictly necessary—since the repository will be downloaded somewhere on the machine where the workflow runs regardless—it gives you more control over its location. This is particularly useful because, a bit later, we will need to specify the path to the `third_parties.repos` file within the repository.

Actions have parameters, described in their repository, that you can specify using the label `uses` to override their default values. This label is used in the third action (`Setup ROS 2`), which requires specifying the `required-ros-distributions` parameter to indicate the ROS distribution to use, as described at `https://github.com/ros-tooling/setup-ros`. The action `ros-tooling/setup-ros` install in the Ubuntu 24.04 LTS a basic ROS 2 Rolling distribution.

- The last action appears significantly more complex. It is the action that compiles the workspace where your repository has been downloaded. The `ros-tooling/action-ros-ci` action runs `rosdep` to install missing dependencies, `colcon build`, and then `colcon test` on the packages specified in the `package-name` parameter.

```
.github/workflows/rolling.yaml

    - name: build and test
      uses: ros-tooling/action-ros-ci@0.3.15
      with:
        package-name: br2_basics br2_bt_bumpgo br2_bt_patrolling
          br2_fsm_bumpgo_cpp br2_fsm_bumpgo_py br2_navigation
          br2_tf2_detector br2_tiago br2_tracking br2_tracking_msgs
          br2_vff_avoidance
        target-ros2-distro: rolling
        vcs-repo-file-url: ${GITHUB_WORKSPACE}/third_parties.repos
        colcon-defaults: |
          {
            "test": {
              "parallel-workers" : 1
            }
          }
```

The `vcs-repo-file-url` parameter, if specified, performs a `vcs-import` of the dependencies file—a process that should already be familiar to the reader, as we covered it in the first chapters of this book.

Finally, for `colcon test`, the number of threads to be launched for testing is set by assigning 1 to `parallel-workers`. Keep in mind that if you run tests in more than one package that use the same topics, the tests may interfere with each other, so you should apply this precautionary measure.

With this workflow configuration file, every time a push is made or a pull request

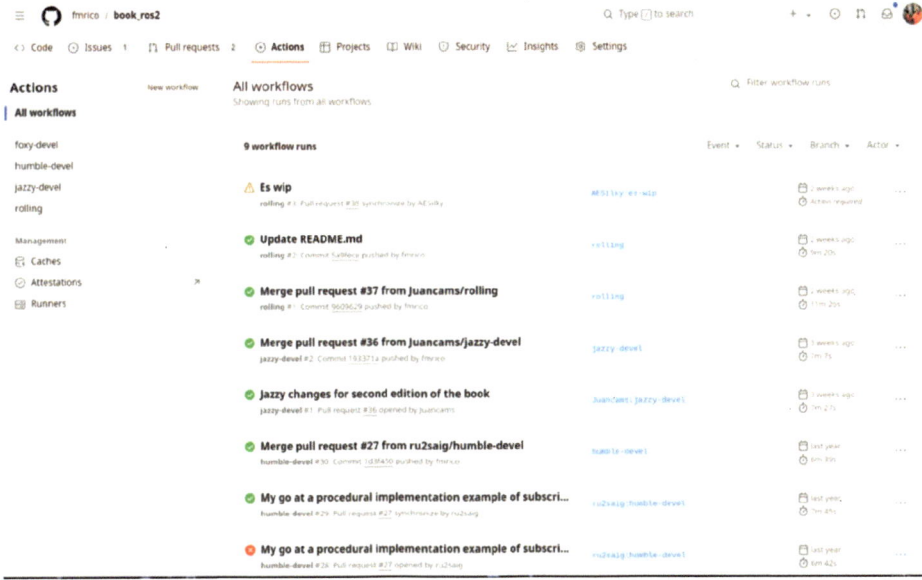

Figure 8.10: Historial of the execution of the action workflows.

is created, this job will run, which can either succeed or fail. If you click on the 'Actions' tab, you can see a history of how each workflow has performed (Figure 8.10). From there, or from wherever the workflow result is displayed, it is easy to access the output of each step in the executed jobs (Figure 8.11). On this screen—or better yet, by accessing the raw logs—you can see why it failed in order to fix the issue. In the *build and test* step, you will see outputs similar to those from running `colcon build` and `colcon test`, along with the log contents.

8.2.4   Repository Structure and Documentation

It is time to focus on the content of your project and how to document it. A repository is not just a place to upload your code; it is the primary site where someone interested in using it will go. It must be well-organized, clear, and provide the necessary information to use it effectively.

Typically, your project will be in a single repository, with all your packages visible from the root directory as separate folders. One recommendation I will make is to separate code packages from interface definitions. This allows you to release binary packages without forcing users who only need the definition of a message—for example, to publish to your nodes' topics—to compile or install your programs they will never execute, along with all the cascading dependencies this might entail. Anyone who has had to cross-compile for another platform knows this well: *Why do I need to compile this library if I only want my robot to understand the messages produced by this software?*.

Regarding naming conventions, these packages were previously named *_msgs, but there is now a trend to name them *_interfaces, as they may contain definitions for services and actions, not just messages.

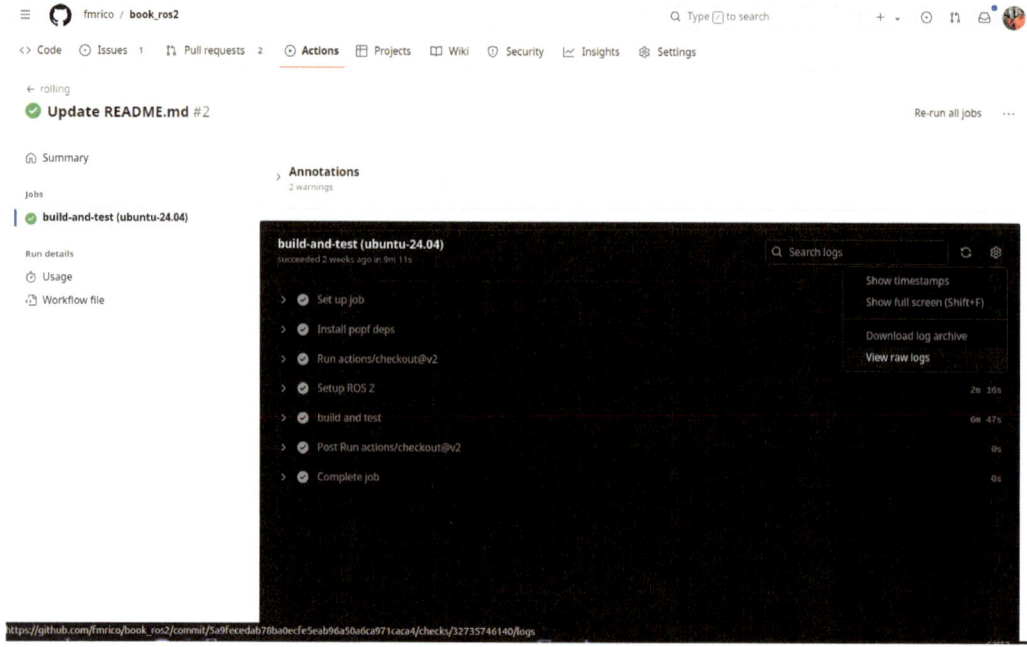

Figure 8.11: Output of the execution of a workflow.

If there are packages you consider optional or do not want to compile and/or release together, you can separate them into multiple repositories and organize everything under a dedicated organization for your project. Let me provide two examples:

- In PlanSys2, there is a GitHub organization(`https://github.com/PlanSys2`) because I want to keep the planning framework packages separate from the examples. Additionally, some plugins (in this case, planners) are not part of the core framework and will not be as well-maintained as the others.

- In MOCAP4ROS2 (`https://github.com/MOCAP4ROS2-Project`), there is a repository for each driver of a motion capture system. These drivers often depend on manufacturer libraries, and maintaining all the drivers in a single repository would force a user interested only in Optitrack, for example, to install Vicon drivers just to compile the repository.

Focusing on your repository, it is a good idea to review the Community Standards page in the Insights tab (Figure 8.12 shows the one for PlanSys2). There are several items listed, and if they are not marked as complete, it will guide you on how to address them. Often, it is as simple as adding a markdown file for the code of conduct or templates for issues and pull requests, or including a file with the license.

What is non-negotiable, however, is having a good `README.md` file, at least at the root of your repository. You can also have a `README.md` for each package if you feel it is clearer to document the nodes and programs individually. However, a general introduction to your project in a root `README.md` is essential.

The first thing that should appear is the repository name and badges that provide

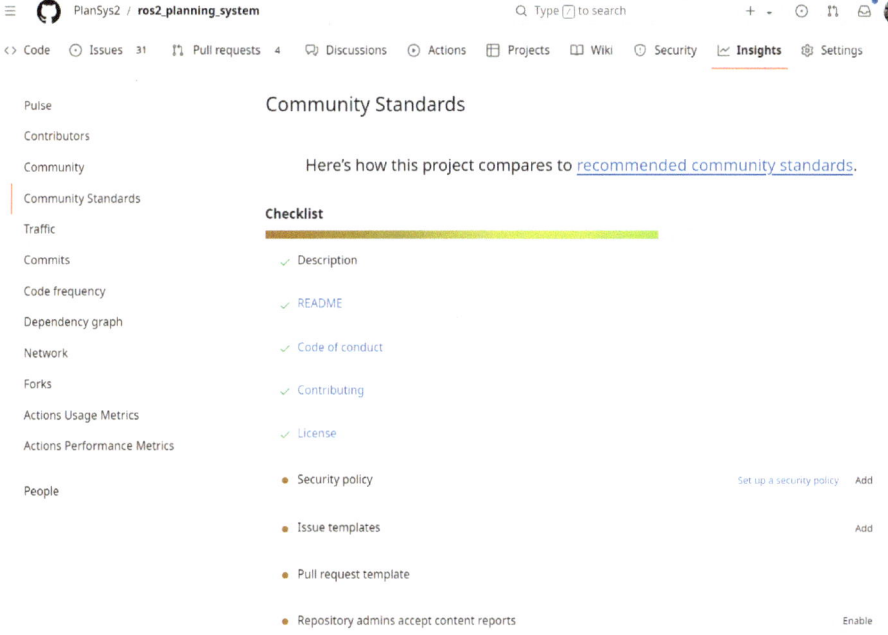

Figure 8.12: Community Standards interface in one GitHub repository.

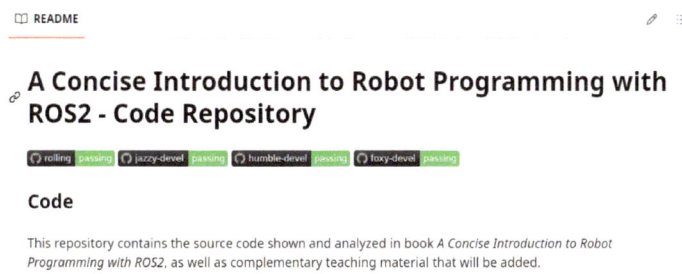

Figure 8.13: Badges in the `README.md` at this book's repository.

information about at least the status of the most recent workflows, as shown in Figure 8.13. To get the text you need to include in the README.md, simply go to the workflow and click on *Create Status Badge,* as shown in Figure 8.14.

You can go a step further and create a meticulously crafted `README.md` like the one for the YASMIN (Yet Another State MachINe) project at `https://github.com/uleroboticsgroup/yasmin`. You can draw many ideas from it, as its structure is very relevant:

- Title.

- Logo (optional, but cool).

- Brief description. One or two lines in enough at the moment.

- CI Badges.

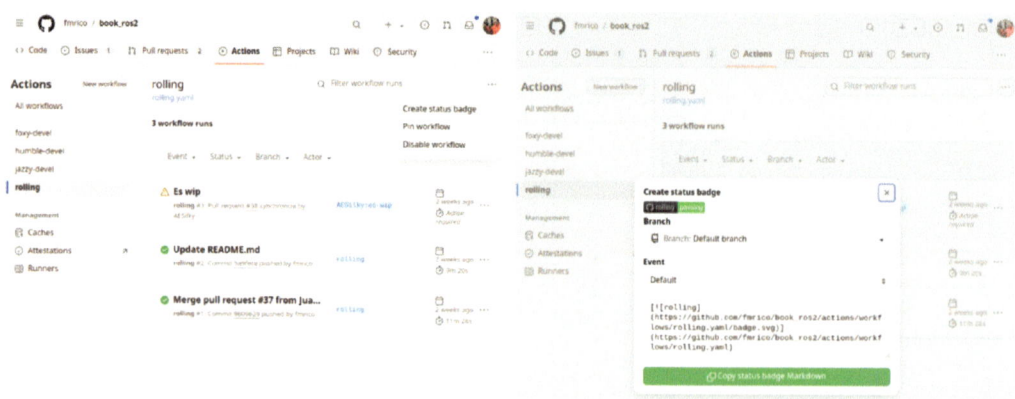

Figure 8.14: Steps to create the badges in markdown.

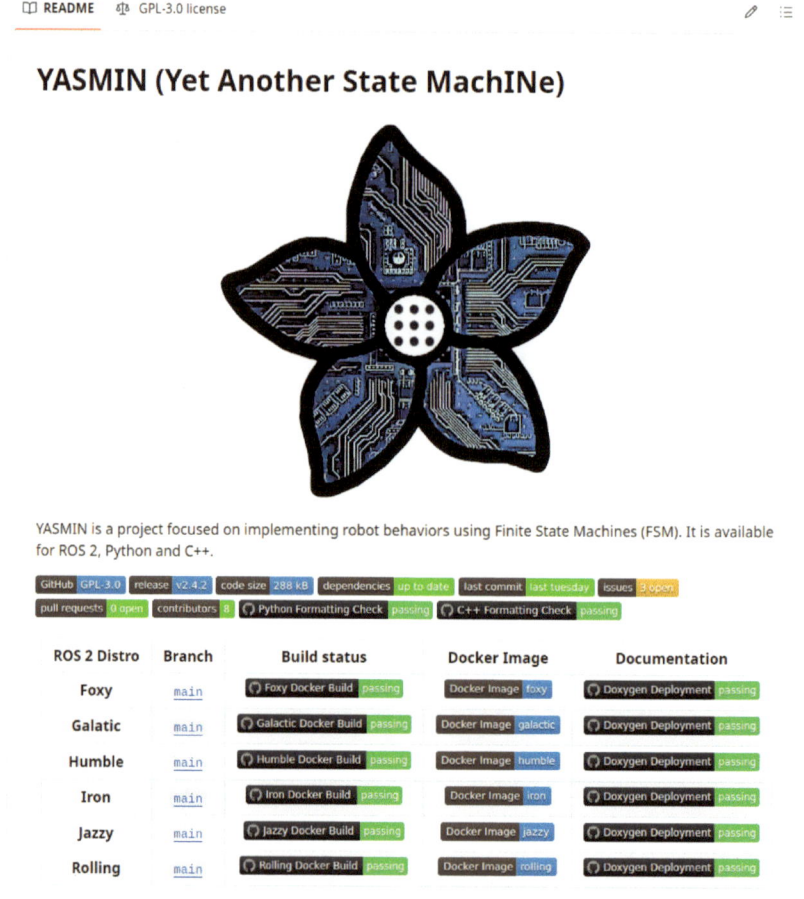

Figure 8.15: Badges in the README.md at YASMIN repository.

- Table of contents.

- General description of the project and its features.

- Installation instructions, both from source and using Docker.

- Demo and screenshots.

- Usage Instructions.

- Citations.This is particularly relevant for those of us working in academia. It is highly recommended to publish scientific papers about your project, if it is possible. A scientific paper allows for a more extensive and formal explanation and analysis than user documentation. You should approach it from a scientific perspective, providing experimental and comparative evidence. The benefit is that those using your software and publishing their results can easily find a way to cite you.

I would recommend adding a description in each package of the nodes it contains, along with a complete description of their parameters, publishers, subscribers, services, and actions. I understand that this can be tedious, but you do not have to forgot tools like ChatGPT, which can automatically generate this documentation for you. Of course, you must carefully review it afterward.

It is increasingly common for complex frameworks to supplement their documentation with a dedicated webpage, often using GitHub Pages and Sphinx. These pages can expand on topics such as getting started, design, tutorials, APIs, or demos, embedding images and videos like those shown in Figure 8.16.

An advantage of this approach is that the language in which the pages are written, reStructuredText (reST), is relatively simple and standard. It is reasonable to ask a contributor to your project to open a pull request on the Sphinx documentation page when proposing a contribution that affects or could expand the existing documentation.

I recommend that anyone wanting to add this type of documentation take a repository like `https://github.com/PlanSys2/PlanSys2.github.io`, duplicate it, and carefully configure the repository—especially the GitHub Pages settings—and then adapt it by changing the text, logo, and whatever else is needed. Whenever you want to regenerate the page, you simply need to run `make html` and `make publish` in the terminal.

8.2.5   Binary Packages Creation

If your project is mature, useful, and you believe it adds value to the ROS ecosystem, it is time to contribute it to a ROS distribution with binary packages. The process is not complicated but does require some work the first time and careful attention.

Much of the complexity is abstracted away by the `bloom`[11] tool, which handles many tasks for you. The ultimate goal of this process is to generate a release for each supported operating system in a release repository and ensure that the `distribution.yaml`[12] file in the `rosdistro` repository[13] includes an entry with your project's information, like the next entry:

---

[11]`https://bloom.readthedocs.io/en/0.5.10/`

[12]`https://github.com/ros/rosdistro/blob/master/jazzy/distribution.yaml`

[13]`https://github.com/ros/rosdistro`

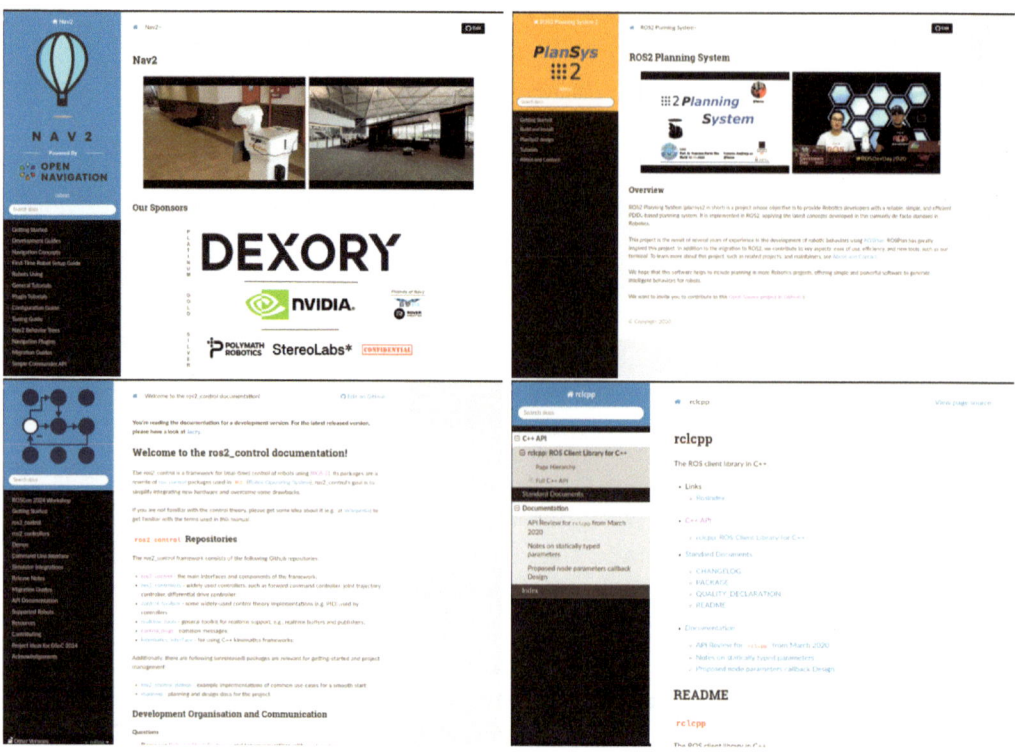

Figure 8.16: Four ROS 2 projects that use GitHub pages + Sphinx approach to complete their documentation.

```
https://github.com/ros/rosdistro/blob/master/jazzy/distribution.yaml

    cascade_lifecycle:
    doc:
      type: git
      url: https://github.com/fmrico/cascade_lifecycle.git
      version: jazzy-devel
    release:
      packages:
      - cascade_lifecycle_msgs
      - rclcpp_cascade_lifecycle
      tags:
        release: release/jazzy/{package}/{version}
      url: https://github.com/ros2-gbp/cascade_lifecycle-release.git
      version: 2.0.0-3
    source:
      type: git
      url: https://github.com/fmrico/cascade_lifecycle.git
      version: jazzy-devel
    status: maintained
```

This file contains all the packages that are part of a distribution so that, during the next synchronization of the distribution, the ROS build farm[14] picks up this file and generates all the binary packages, including yours. The process described here will conclude with a pull request to the `rosdistro` repository to include this entry, which must be approved by one of the ROS bosses of the distribution.

---

[14]https://build.ros.org/

In this section, we will describe step by step the process of creating binary packages for a project that has not been previously released[15] .For this purpose, I will use the project created in the previous chapter, YAETS, which is available at `https://github.com/fmrico/yaets` and has not been released at the time of writing these lines. This is quite fortunate, as I will be able to release it while writing, ensuring that the instructions are complete. It is a win-win!

We start with a repository containing a single package, `yaets`, with two branches: `rolling` and `jazzy-devel`. It includes tests and well-written documentation in the README.md, complete with CI status badges. The project has a OSI Approved license[16], Apache 2.0, in this case, and the name compliants with REP 144[17]. So, we are ready to realease.

We are going to create the packages for the Jazzy distribution, starting from the `jazzy-devel` branch. Currently, this branch is synchronized with the `rolling` branch, as I just created the `jazzy-devel` branch from the `rolling` branch. Since Jazzy is the latest version, it recently branched off from Rolling, so it sounds reasonable. Let's look at the following steps:

1. First, we need to create a separate repository where the various releases will be generated. The `bloom` tool will manage this for you. Under normal circumstances, you will not need to make any manual commits to this repository. There are two options for release repositories:

   (a) This has been the usual option until very recently. It will be the one we use in this section, as it is the most straightforward for completing the process directly. Only allowed for stable distributions. You could not release packages for Rolling using this approach.

      Typically, if the project's repository is `https://github.com/fmrico/yaets`, the releases repository would be `https://github.com/fmrico/yaets-release`, as shown in Figure 8.17.

   (b) The documentation[18] recommends that release repositories be hosted under the GitHub organization `https://github.com/ros2-gbp`. The process is not straightforward, as you must request the creation of the repository via a pull request. Additionally, you need to request the creation of a release team that includes your GitHub user to manage this release repository. This is the recommended option and may soon become the only option.

   The advantage of this process is that you can currently choose the first option (1a), which we will use in this section, and simultaneously request the creation of

---

[15]You can also find information of this process at `https://docs.ros.org/en/rolling/How-To-Guides/Releasing/Releasing-a-Package.html`

[16]`https://opensource.org/licenses`

[17]`https://www.ros.org/reps/rep-0144.html`

[18]`https://docs.ros.org/en/rolling/How-To-Guides/Releasing/Release-Team-Repository.html#create-a-new-release-repository`

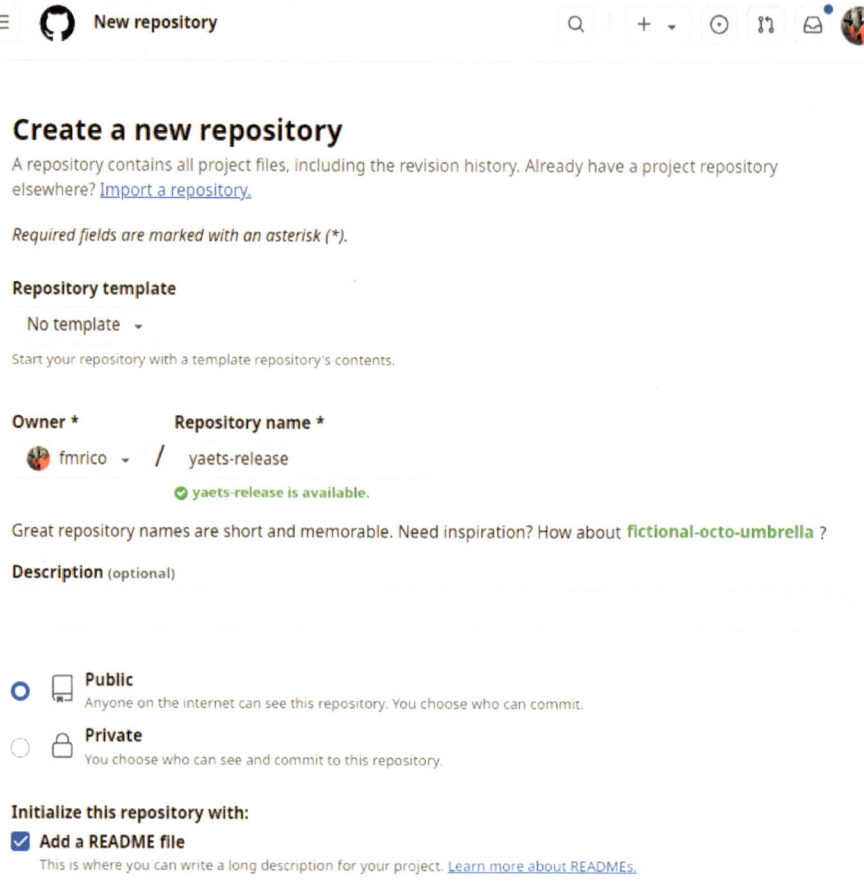

Figure 8.17: Creation of the release repository.

your release repository, specifying that an external one already exists. Through this process, the release repository will eventually migrate to the `ros2-gbp` organization.

2. On my local machine, I open a terminal, navigate to the directory where I have YAETS cloned, and ensure I am on the `jazzy-devel` branch. The first step is to generate the changelog for the new version to be released:

```
$ cd src/yaets
(jazzy-devel)$ catkin_generate_changelog --all
Found packages: yaets
Querying all tags and commit information...
Generating changelog files with all versions...
- creating './CHANGELOG.rst'
Done.
Please review the extracted commit messages and
consolidate the changelog entries before committing the files!
```

3. As indicated by the output of the previous command, the next step is to clean the changelog to remove commit messages related to minor changes, duplicates,

management tasks, etc. In other words, remove the entries that do not reflect the actual progress made in your project for this release. For example, Figure 8.18 shows the original state of the file on the left and the cleaned version on the right.

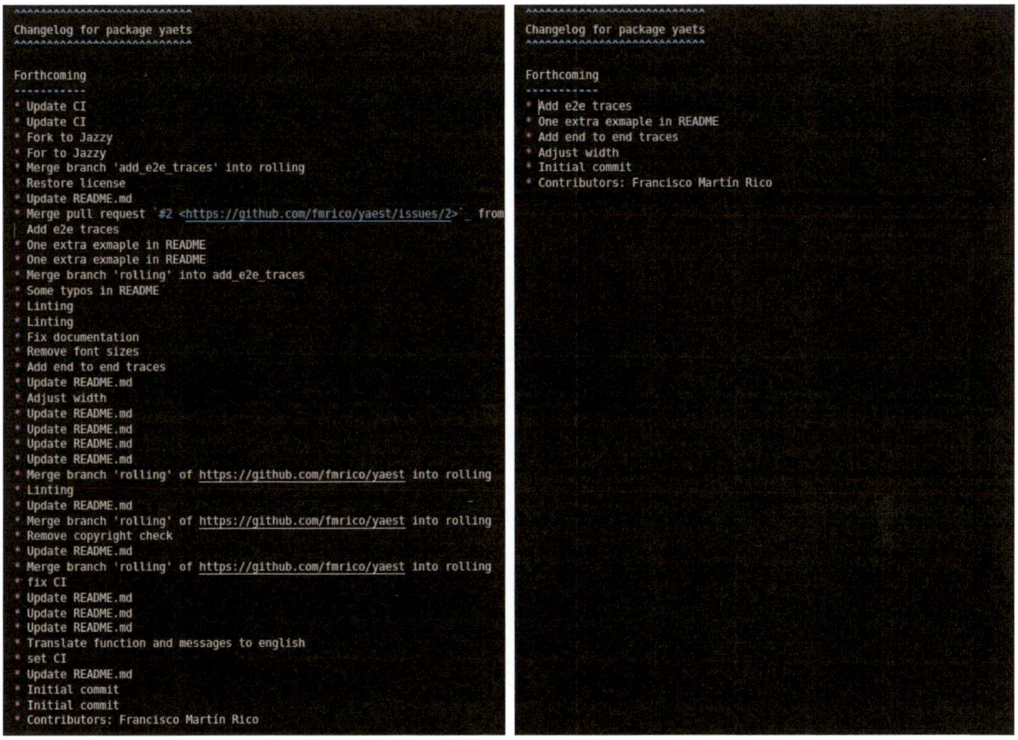

Figure 8.18: Cleaning the Changelog.

This step also includes committing the modified file. This is important because, otherwise, the process will fail.

```
(jazzy-devel)$ git add CHANGELOG.rst
(jazzy-devel)$ git commit -s -m "Update Changelog"
```

4. In the next step, the version will be incremented, as it is required that the new version is higher than the one already in the repository. The following command will update the version in the package.xml, replace the Forthcoming string in the changelog with the new version and its date, and finalize this release in the project's repository:

```
(jazzy-devel)$ catkin_prepare_release
```

When it asked me if I wanted to upload the changes, I responded 'yes.' Typically, the default option for these questions is quite safe if you are unsure of the correct answer.

5. In this final step, the releases will be created and added to the release repository, concluding with a pull request to the rosdistro repository. There is a difference

between releasing a package for the first time and releasing a new version of the same package. The first time, you will be asked several questions, while subsequent releases will be simpler. The process involves a single command:

```
(jazzy-devel)$ bloom-release --new-track --rosdistro jazzy --track jazzy
yaets
```

During this process, several questions will be asked, and it is important to answer them correctly:

- Release repository is: `https://github.com/fmrico/yaets-release.git`.
- Repository Name: yaets.
- Upstream Repository URI: `https://github.com/fmrico/yaets.git`
- For Upstream VCS Type, Version and Release Tag I accepted the default values.
- Upstream Devel Branch: `jazzy-devel`
- For ROS Distro, Patches Directory, and Release Repository Push URL I accepted the default values.
- When asked if I want to add documentation information for this repository, I said yes, and accepted the default options which, basically, are pointing to the documentation existing in the repository.
- Finally, it asks if it should create the pull request to rosdistro for us. I respond 'yes,' and the result is this pull request (Figure 8.19). Once it is accepted, it will mean that the package will officially be part of the Jazzy distribution of ROS. Yay!

To release new versions, the entire process will need to be repeated exactly as described. Naturally, the release repository will already exist, and the command will now be:

```
(jazzy-devel)$ bloom-release --rosdistro jazzy yaets
```

It is common, especially the first few times, for everything not to go smoothly in the build farm. Even if your repository has passed its CI checks, the build farm might detect errors that did not appear earlier. A common issue is forgetting to include a dependency in the `package.xml`. You will receive an email from the ROS 2 Buildfarm indicating that the 'Build failed in Jenkins.' Go to the link provided in the email, check the error, and work on resolving it. The ROS bosses may be able to assist you. Once resolved, you will have no choice but to make a new release with the fix. I have personally had occasions where fixing an error required making releases over several days until the ROS 2 Buildfarm finally stopped reporting issues.

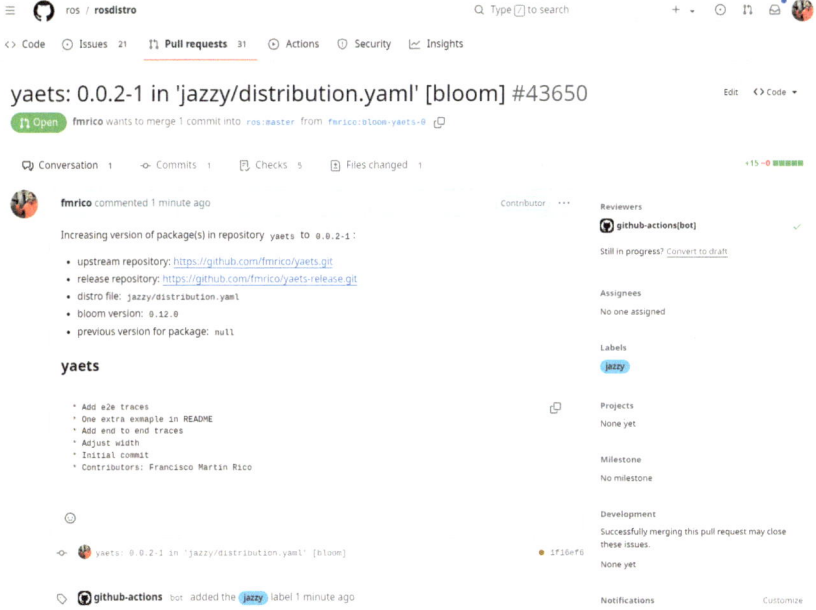

Figure 8.19: Pull request to the rosdistro repository.

8.2.6 Tips and Best Practices for Managing Contributions

When your project starts receiving more and more contributions, you find yourself spending less time coding and testing new ideas and more time interacting with contributors, responding to issues, and reviewing pull requests. This is, on one hand, very rewarding, but on the other, it can become a workload that risks burning you out. This section offers some advice to help you as you begin to take on the responsibilities of maintaining your own project.

- **Documenting your project is essential**. The gaps you leave will be filled by contributors based on their own expectations, desires, or requirements. Do not leave ambiguities when describing the purpose of your project and defining the scope of contributions you are willing to accept. This does not mean you cannot encourage discussions with contributors about the next steps for your project, but it helps you avoid having to say 'no' too often when someone makes a contribution that does not align with your vision.

  It is useful to write a brief list of bullet points outlining the types of contributions you are open to and the conditions for accepting them. While it may not be necessary to state that 'pull requests must pass style tests,' it can be helpful to mention that contributions should not reduce test coverage or that new functionalities must be documented in specific ways. Setting these expectations in advance will help prevent misunderstandings. You can also manage expectations regarding your availability by clearly stating how much time you can realistically dedicate to a project.

- One of the wonderful aspects of open source is that **you contribute to your project freely**, with no obligations other than those you choose to impose on yourself out of responsibility to your users. No one can demand that you work more, review their contributions faster, or similar requests. If they want to make demands, it is because they are prepared to pay for it; otherwise, they must adapt to your pace and your vision

"You don't get it. I built this place. Down here I make the rules. Down here I make the threats. Down here, I'm God". Trainman, Matrix Revolutions, 2003.

- **Discussions about a project should be public**, documented, and accessible to both current and future contributors. If a contributor contacts you privately via email to discuss potential contributions or technical decisions, you should immediately encourage them to open an issue where the topic can be discussed publicly. It is essential to ensure that no contributor has access to more information than others.

- **Be friendly but firm**. Sometimes, you need to know how to say no. It is not fun, but it is necessary. There will always be people who, despite your best efforts to document your project and clearly outline the types of contributions that fit, have not taken the time to read the documentation. Some contributions simply will not align with your vision for the project, and this must be respected. However, it is important to communicate this in a friendly and constructive manner. After all, the person contributing is using your software and has dedicated their time and effort to create something they believe is helpful. They deserve your gratitude, feedback, and suggestions for alternatives, but you are not obligated to accept their proposal. Be friendly and constructive, as I was with the contribution shown in Figure 8.20, where I tried to offer an alternative while declining the original contribution.

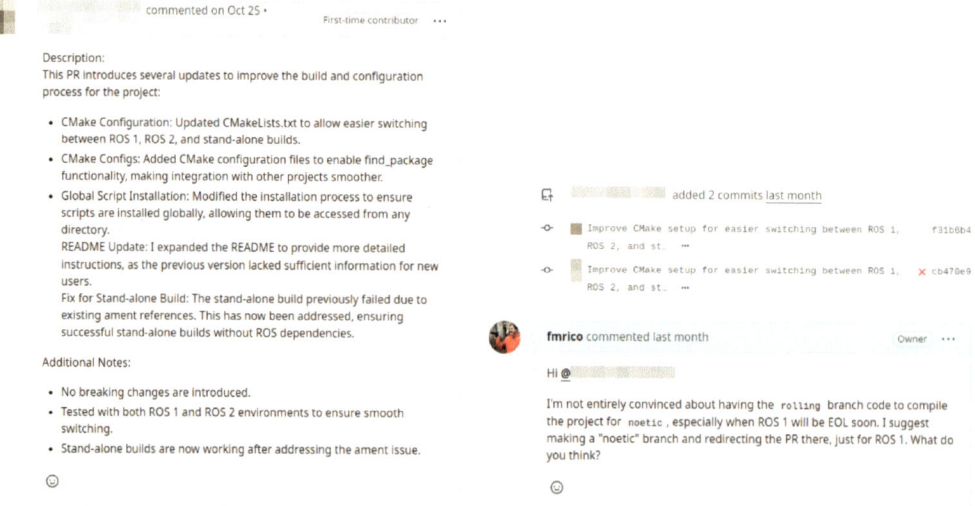

Figure 8.20: Gentle conversation at the YAETS repository.

- If you notice a user is enthusiastic about the project but could use some refinement in their approach or skills, take the time to **help them level up**. No one is born knowing everything, and investing a bit of time in mentoring them can be more productive for you in the long run and help avoid frustration for both sides.

- Do not feel frustrated or upset **if someone forks your project** and begins developing new features independently. It is likely that their needs differ significantly from the goals of your project. In such cases, this approach is often better than having a contributor reluctantly steer your project in a direction that aligns only with their specific use case.

- Make full **use of all the automated tools at your disposal**: GitHub Actions, templates, test coverage reports, and more. These resources can streamline your workflow, ensure consistency, and reduce the manual effort required to maintain your project effectively.

- Finally, there is nothing wrong with **taking a break** from your project if you start noticing signs of burnout. You must be happy with your project. Let others know you will be stepping away for a while, and return to it when you feel ready. Alternatively, consider reaching out to one of the main contributors to see if they are willing to take over maintaining the project in your absence, or maybe forever.

Well, we have reached the end of this book. I hope you enjoyed reading it as much as I enjoyed writing it. I look forward to crossing paths with you again—whether on ROS Discourse, in an issue, or in person at a ROSCon. If we meet in person, it would make me very happy if you came to say hello, shared what you liked or did not like about this book, and we could discuss it over a drink.

As we used to say, **GO ROS!!!**

Create a Repository

From scratch - Create a new local repository

```
$ git init [project name]
```

Download from and existing repository

```
$ git clone my_url
```

Download from and existing repository an specific branch

```
$ git clone -b my_branch my_url
```

Observe the Repository

List new or modified files not yet committed

```
$ git status
```

Show the changes to files not yet staged

```
$ git diff
```

Show the changes to staged files

```
$ git diff --cached
```

List the change dates and authors for a file

```
$ git blame [files]
```

Show change history in fancy way

```
$ git log --oneline --decorate
--graph --all
```

Show change history

```
$ git log
```

Show change history for file/directory including diffs

```
$ git diff HEAD
```

Show the changes between two commits

```
$ git diff commit1 commit2
```

Working with Branches

List all local branches

```
$ git branch
```

List all branches, local and remote

```
$ git branch -av
```

Switch to a branch and update working directory

```
$ git checkout my_branch
```

Create a new branch from current one

```
$ git checkout -b new_branch
```

Delete a branch LOCALLY

```
$ git branch -d my_branch
```

Delete a branch REMOTELY

```
$ git push --delete origin my_branch
```

Merge branch_to to branch_from

```
$ git checkout branch_to
$ git merge branch_from
```

Working with remotes

Check remotes

```
$ git remote -v
```

Add remote repository

```
$ git remote add [name] my_url
```

Stashing

Store unstaged changes for later

```
$ git stash
```

Apply stored changes

```
$ git stash pop
```

Make a change

Stage files, ready for commit

```
$ git add [files]
```

Stage all changed files, ready for commit

```
$ git add --all
```

Commit all staged files with signed history

```
$ git commit -s -m "commit message"
```

Unstage files, keeping file changes

```
$ git reset [files]
```

Revert everything to the last commit

```
$ git reset --hard
```

Synchronize

Get the Latest changes from origin (no merge)

```
$ git fetch
```

Get the Latest changes from origin and merge

```
$ git pull
```

Get the Latest changes from origin and rebase

```
$ git pull --rebase
```

Push local changes to origin

```
$ git push
```

Push local new branch to origin

```
$ git push -u origin new_branch
```

Git cheat sheet for the book "A Concise Introduction to Robot Programming with ROS2" 2nd Edition. Francisco Martín Rico. CRC Press 2025

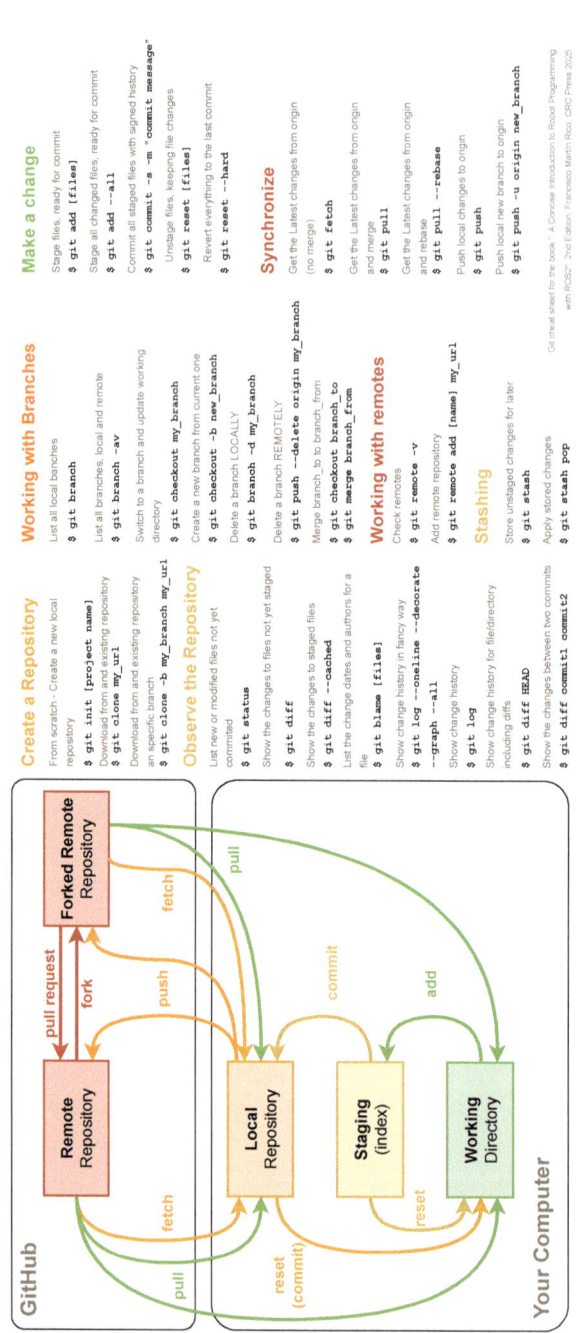

Figure 8.21: Git cheatsheet.

Bibliography

[1] S. Macenski, T. Foote, B. Gerkey, C. Lalancette, and W. Woodall, "Robot operating system 2: Design, architecture, and uses in the wild," *Science Robotics*, vol. 7, no. 66, p. eabm6074, 2022.

[2] G. Metta, P. Fitzpatrick, and L. Natale, "Yarp: Yet another robot platform," *International Journal of Advanced Robotic Systems*, vol. 3, no. 1, p. 8, 2006.

[3] M. Montemerlo, N. Roy, and S. Thrun, "Perspectives on standardization in mobile robot programming: The carnegie mellon navigation (carmen) toolkit," in *Proceedings 2003 IEEE/RSJ International Conference on Intelligent Robots and Systems (IROS 2003)(Cat. No. 03CH37453)*, vol. 3, pp. 2436–2441, IEEE, 2003.

[4] B. Gerkey, R. T. Vaughan, A. Howard, *et al.*, "The player/stage project: Tools for multi-robot and distributed sensor systems," in *Proceedings of the 11th international conference on advanced robotics*, vol. 1, pp. 317–323, Citeseer, 2003.

[5] M. Quigley, B. Gerkey, K. Conley, J. Faust, T. Foote, J. Leibs, E. Berger, R. Wheeler, and A. Ng, "Ros: an open-source robot operating system," in *Proc. of the IEEE Intl. Conf. on Robotics and Automation (ICRA) Workshop on Open Source Robotics*, (Kobe, Japan), May 2009.

[6] S. Thrun, D. Fox, W. Burgard, and F. Dellaert, "Robust monte carlo localization for mobile robots," *Artificial Intelligence*, vol. 128, no. 1, pp. 99–141, 2001.

[7] R. A. Brooks, "Elephants don't play chess," *Robotics and Autonomous Systems*, vol. 6, no. 1, pp. 3–15, 1990. Designing Autonomous Agents.

[8] A. Marzinotto, M. Colledanchise, C. Smith, and P. Ögren, "Towards a unified behavior trees framework for robot control," in *2014 IEEE International Conference on Robotics and Automation (ICRA)*, pp. 5420–5427, 2014.

[9] S. Macenski, F. Martin, R. White, and J. Ginés Clavero, "The marathon 2: A navigation system," in *2020 IEEE/RSJ International Conference on Intelligent Robots and Systems (IROS)*, 2020.

Index